CONDITIONS OF LIFE IN THE SEA

This is a volume in the Arno Press collection

HISTORY OF ECOLOGY

Advisory Editor
Frank N. Egerton III

Editorial Board
John F. Lussenhop
Robert P. McIntosh

See last pages of this volume for a complete list of titles.

CONDITIONS OF LIFE
IN THE SEA

JAMES JOHNSTONE

ARNO PRESS
A New York Times Company
New York / 1977

Editorial Supervision: LUCILLE MAIORCA

Reprint Edition 1977 by Arno Press Inc.

First published 1908
Reprinted by permission of Cambridge
University Press

Reprinted from a copy in
The University of Illinois Library

HISTORY OF ECOLOGY
ISBN for complete set: 0-405-10369-7
See last pages of this volume for titles.

Manufactured in the United States of America

Publisher's Note: The frontispiece has been
reproduced in black and white in this edition.

Library of Congress Cataloging in Publication Data

Johnstone, James, 1870-1932.
   Conditions of life in the sea.

   (History of ecology)
   Reprint of the 1908 ed. published by Cambridge
University Press, Cambridge, Eng., in series:
Cambridge biological series.
   Bibliography: p.
   1. Marine biology. I. Title. II. Series.
QH91.J58 1977   574.92   77-74232
ISBN 0-405-10401-4

CAMBRIDGE BIOLOGICAL SERIES
GENERAL EDITOR :—ARTHUR E. SHIPLEY, M.A., F.R.S.
FELLOW AND TUTOR OF CHRIST'S COLLEGE, CAMBRIDGE

CONDITIONS OF LIFE
IN THE SEA

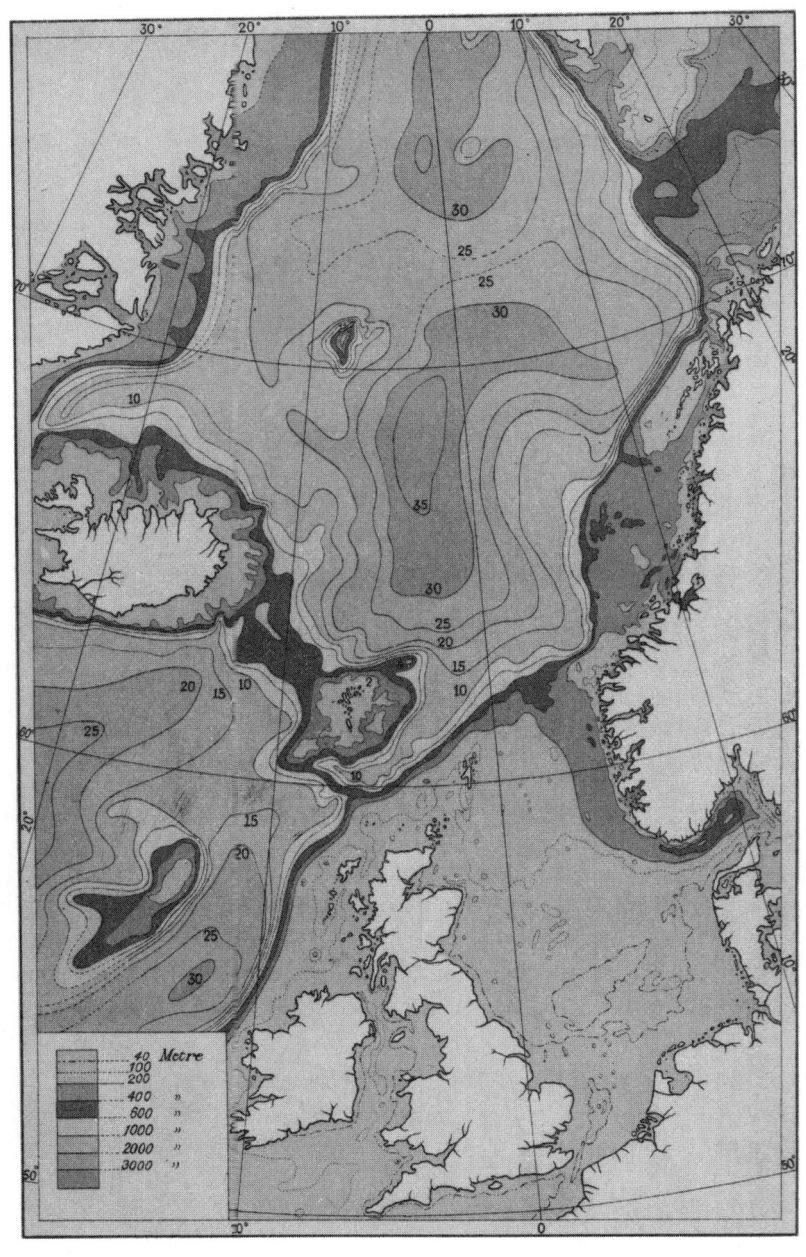

THE BRITISH PLATEAU AND NORWEGIAN SEA.

(From *Rapports et Procès-Verbaux; International Fishery Investigations.*)

CONDITIONS OF LIFE IN THE SEA

A SHORT ACCOUNT OF QUANTITATIVE MARINE BIOLOGICAL RESEARCH

BY

JAMES JOHNSTONE

FISHERIES LABORATORY, UNIVERSITY OF LIVERPOOL

CAMBRIDGE:
at the University Press
1908

Cambridge:
PRINTED BY JOHN CLAY, M.A.
AT THE UNIVERSITY PRESS.

[*All rights reserved*]

PREFACE

THE object of this book is to give a short account of the main results of modern quantitative marine biological investigations, and of the related results of hydrography and oceanography. There is at present no adequate summary of these researches in English; and although several German accounts have been published these do not seem to be very well known in this country. Already a great number of memoirs have been written, but the publications in which these have appeared are not very accessible, even in the ordinary biological libraries. Much of the work, especially that relating to the methods of plankton research, and to the hydrodynamical investigation of ocean currents, is very technical and can hardly be summarised in a book of this character. So I have referred to the most important memoirs, and from these, and the further references which they contain, the reader will be able to tap the original sources of information. Part I. of the book is rather elementary in treatment, the object being to supply an account of those facts of oceanography which are not likely to be familiar to the reader who is not specially interested in marine biological investigation. Part II. deals with the methods and results of quantitative marine biological research; and Part III. with the general conditions of life in the sea.

It is characteristic of a really great idea in science that it should stimulate further discovery by the suggestion of new lines of research, and new methods of investigation. Hensen's primary object in devising methods and apparatus by means of which it might be possible to estimate the actual numbers of the microscopic animals and plants inhabiting parts of the sea, seems to have been the calculation of the numbers of certain species of edible fishes in the North Sea and Baltic. But whenever it became possible to estimate (even roughly) the numbers of plants and animals, of microscopic dimensions, in restricted sea-areas, a host of interesting questions immediately suggested themselves; and these would not have arisen—indeed did not arise—apart from the stimulus of the application of methods of quantitative research to marine biological investigation. Thus we had Hensen's discussions of the productivity of the sea, and his comparison of this with the fruitfulness of the land; and Brandt's investigation of the chemical composition of the plankton, with all the interesting questions and speculations which were suggested by the results of these two series of researches. Then whenever the plankton became regarded as the "pastures of the sea" the question of the distribution of the sources of nutriment of these marine pastures inevitably suggested itself; and new and accurate determinations of the proportions and distribution of the ultimate inorganic food-stuffs of marine organisms were instituted. And when it was found by means of quantitative plankton investigations that the mass of life in different parts of the oceans of the world varied greatly per unit volume of sea, it was seen that the cause of this variation could reside only in the concomitant variation of the proportions of the ultimate inorganic food-stuffs. But why do the latter vary? Brandt suggested bacterial action, and very soon the investigation of the nitrogen-bacteria of the sea commenced—with the most fruitful results—and a further insight was gained into the question of the circulation of nitrogen in nature. Other causes were then suggested, and it has become probable that the circulation of food-stuffs in the sea is to be closely associated with the distribution of oceanic streams and currents.

Modern oceanographical research has been developed from the work of Murray, Buchanan and Dittmar, which was based on the collections and observations made during the *Challenger* expedition. Cleve and Pettersson then attempted to trace the origin of the water which, from season to season, drifted to the shores of Scandinavia. Knowing that different sea-areas were characterised by different *facies* of microscopic life they endeavoured to press the study of the plankton into the service of hydrography, and hydrography into that of meteorology. Like Hensen's, their object was a restricted one, but the idea involved was so fruitful that a host of investigations sprang from it. To it we owe new determinations of the chemical composition, and the physical properties of sea water; and the invention of beautiful apparatus for the tracing and measurement of ocean currents; as well as the mathematical investigations by Sandström, Helland-Hansen, and D'Arcy Thompson of the movements of water-masses in the sea. Then if different plankton types are to be associated with water of different chemical and physical properties the causes are to be sought. Again new investigations were instituted, and the study of the association of changes in the salinity and temperature of the sea, with concomitant changes in the nature and abundance of the plankton, was extended to the study of the migrations and breeding habits of fishes. Queer side-issues have arisen in the course of these investigations: the study of the physical condition of the particles composing the sea-bottom, for instance, and the association of changes in the viscosity of the sea water with the movements of the plankton.

Possibly more was anticipated from the direct results of the quantitative investigation of the plankton than has actually been acquired. Probably too, Hensen, Cleve and Pettersson under-rated the difficulties of the work, and under-estimated or overlooked sources of error. It would be strange indeed if they did not! Criticism, both of methods and results, has not been wanting. How much, for instance, has the quantitative study of marine life been prejudiced in this country by the publication of Haeckel's *Plankton-Studien*! And how many people who have read that brilliant but injudicial essay have seen Hensen's reply to it? I have tried critically to consider the weight of the various objections that have been made against conclusions deduced from the quantitative study of marine life, and trust that I have not under-estimated these. Adverse criticism, which would surely have been fatal to the development of an unsound method of investigation, has certainly stimulated research on the lines suggested by Hensen. Lohmann's work, for instance, was essentially critical and yet, more than any other piece of plankton investigation published since the appearance of Hensen's first memoir, it will facilitate the quantitative study of the plankton. Admittedly imperfect and incomplete as are these investigations, their results form nevertheless the most fascinating chapter of modern marine biological research, and as such may be commended to the reader.

I am indebted to my colleagues Miss M. Allen, and Mr H. J Buchanan-Wollaston for much kind assistance; to Dr F. W. Gamble for reading part of the manuscript; and to Mr A. E. Shipley for many suggestions, and much advice.

J. J.

LIVERPOOL,
July 1908.

CONTENTS

PART I.

INTRODUCTORY.

CHAPTER I.

THE EXPLORATION OF THE SEA.

Methods of oceanographical investigation—The Challenger—Later Expeditions—Oceanographical apparatus—Sounding machines—Water-sample bottles and thermometers—Salinity and density of sea water—The International Fishery Investigations—Biological investigations—Dredging and trawling—Beam trawl and otter trawl—Deep sea dredging and trawling—Other fishing apparatus—Pelagic fishing apparatus—Tow-nets and similar apparatus . . . . Page 1

CHAPTER II.

THE OCEANOGRAPHY OF THE NORTH-WESTERN OCEAN.

The British Islands and the continental shelf—The Greenland-Shetland elevation—Norwegian and Atlantic basins—The fore-shore—Sea-bottom deposits—Terrigenous deposits—Pteropod Ooze—Globigerina Ooze—Radiolarian Ooze—Diatom Ooze—Red clay—Temperature of the sea—Colour and transparency of the sea—The water circulation in the North Atlantic and Polar seas—The Gulf Stream and its offshoots—The European Stream—The Wyville-Thompson barrier—The Norwegian branch of the European Stream—Perturbations and periodic variations of the European Stream—Climate and hydrography 30

CHAPTER III.

LIFE IN THE SEA.

Zones of life—Littoral, laminarian and coralline zones—Benthos, nekton and plankton—The deep-sea fauna—The migratory fauna—The plankton—Constituents of the plankton, Fishes, Ascidians, Crustacea, Mollusca, Worms, Echinoderms, Coelenterates, Ctenophores, Protista, Peridinians, Foraminifera, Radiolaria, Diatoms, other Protista **54**

CHAPTER IV.

LIFE IN THE SEA.

Permanent and transitory plankton—Life-histories of marine animals—Direct and indirect modes of development—Larvae of marine animals—The struggle for existence in the sea—Destruction in the sea—Fecundity of marine animals—Protection of the eggs and larvae—Seasonal variations in the plankton—Migrations of fishes—The struggle for food in the sea . . . **81**

CHAPTER V.

THE SEA-FISHERIES.

Economic marine products—Exploitation of the sea—Influence of man on the marine fauna—Methods of fishing—Trawlers, drifters and liners—Pelagic and demersal fishes—Classes of fishermen and their economic condition—The shell-fisheries **102**

PART II.

QUANTITATIVE MARINE BIOLOGY.

CHAPTER VI.

QUANTITATIVE PLANKTON INVESTIGATION.

The Hensen nets and methods—Construction and method of use of quantitative plankton nets—The catch and its preservation—The "Korbnetz"—The Petersen-Hensen net—Estimation of the catch of a quantitative plankton net—Volume, chemical composition, mass and enumeration of the catch—The Stempel-pipette—Validity of the Hensen methods—The objections of Haeckel and Kofoid—Lohmann's investigations—Other quantitative plankton methods—Capture of plankton by appendicularians—Defects of the Hensen methods 117

CHAPTER VII.

THE DISTRIBUTION OF THE PLANKTON.

The mixing of the plankton in the sea—Experimental errors in the Hensen methods—Vertical distribution of the plankton—Stratification of the plankton —Heliotropic migrations—Changes in the distribution of the plankton due to temperature, viscosity and convection currents—Inshore and offshore plankton —Plankton types—Oceanic and neritic plankton—Distribution of the plankton according to hydrographic conditions—Atlantic and North Sea plankton— Cleve's investigations—Plankton mixtures—Halistatic planktons . 142

CHAPTER VIII

A CENSUS OF THE SEA.

Difficulties and uncertainties of the estimation of the life of the sea— Density of the plankton of the North Sea—Average plankton contents of the North Sea—Abundance of diatoms, peridinians and copepods in the sea— Variations in the density of the plankton with the locality—Density of the micro-plankton—Density of fish-eggs and larvae in the sea—Hensen's "Nordsee Expedition" of 1895—Absolute abundance of certain species of fish in the North Sea—Errors in the estimations—Density of life on the sea bottom 157

CHAPTER IX.

THE PRODUCTIVITY OF THE SEA.

Productivity of the North Sea—Produce of the North-Western fishing area —Productivity of an inshore fishing area—Productivity of the shell-fisheries of the West coast of England—Cultivated water areas—German carp fisheries —A cultivated mussel fishery—The absolute productivity of a sea area— Composition of marine produce—Relative productivity of the sea and the land —The impoverishment of the sea—Productivity of the sea in different latitudes 178

PART III.

METABOLISM IN THE SEA.

CHAPTER X.

THE CONDITIONS OF LIFE IN THE SEA.

The nutrition of marine organisms—Plankton as food-stuff—Ultimate food-stuffs in the sea—Modes of nutrition among marine organisms—Holophytic organisms—Saprophytic organisms—Myxotrophic animals—Saprozoic animals—Law of the minimum—Physical conditions and the metabolism of marine organisms—Temperature—Salinity—Migrations of fishes and salinity changes—Sunlight—Photo-synthesis—Colour of sea water . . . **206**

CHAPTER XI.

BACTERIA IN THE SEA.

Types of bacteria—Biological reactions—Distribution of bacteria in the sea—Halibacteria—Luminous bacteria—Putrefactive bacteria—Prototrophic, metatrophic and paratrophic bacteria—Fermentation bacteria—Sulphur bacteria—Nitrifying bacteria—Nitrogen-fixing bacteria—Denitrifying bacteria—Temperature and denitrification **253**

CHAPTER XII.

THE CIRCULATION OF NITROGEN.

Atmospheric, organic, and inorganic nitrogen—Animal and plant metabolism—Elimination of waste products—Putrefactive decomposition—Breaking down of dead organic matter—Nitrification—Sewage and its purification—Denitrification—Drainage of nitrogen from land to sea—Mass of nitrogen annually added to the sea—Utilisation of land drainage by marine organisms—Transmutation of nitrogen through series of marine organisms—Production by marine plants—Transfer of nitrogen from the sea to the land—Destruction of nitrogen compounds in the sea—Denitrification in the sea—Density of life in polar and tropical seas—Circulation of inorganic food-stuffs—Structural and current metabolism **273**

APPENDIXES

		PAGE
I.	CHEMISTRY OF THE PRIMITIVE OCEAN	299
II.	COMPOSITION OF DEEP-SEA DEPOSITS	304
III.	FOUR HENSEN NET HAULS IN THE IRISH SEA	305
IV.	ACCURACY OF QUANTITATIVE PLANKTON OBSERVATIONS	306
V.	CALCULATION OF THE COEFFICIENTS OF THE QUANTITATIVE PLANKTON NETS	313
VI.	CALCULATION OF THE AGE OF THE EARTH FROM THE RATE OF DRAINAGE OF SALT FROM THE LAND	315
VII.	BIBLIOGRAPHY	316
	INDEX	320

ILLUSTRATIONS

Chart of the British Plateau and Norwegian Sea . . *Frontispiece*

FIG.		PAGE
1.	The Lucas automatic sounding machine	4
2.	The stop-cock sounding lead	6
3.	Deep-sea reversing thermometer, Kiel hydrometer . .	8
4.	Nansen-Pettersson water-bottle	9
5.	International Fishery Investigation area	15
6.	Naturalist's dredge	16
7.	Beam trawl-net	17
8.	Otter trawl-net	19
9.	Arrangement of trawl gear	20
10.	Agassiz trawl-net	23
11.	Sea-bottom fishing basket	24
12.	Ordinary surface tow-net	26
13.	Heligoland "Scherbrutnetz"	27
14.	Sea-bottom deposits	35
15.	Hydrographic section shewing temperature over the Wyville-Thompson Ridge	39
16.	Similar section shewing salinities	40
17.	Chart of North-Western Ocean shewing the European Stream and its offshoots	47
18.	Plankton : fishes, worms, etc.	68
19.	Plankton : crustacea	71
20.	Plankton : protozoa, etc.	76
21.	Plankton : diatoms	79
22.	Succession of fish ova in the plankton of Cardigan Bay, 1905-6	95
23.	Rough succession of plankton in the sea off N.W. England .	96
24.	The vertical quantitative plankton net	122
25.	The Hensen "Filtrator"	125
26.	The Korbnetz	127
27.	The "Stempel-pipette"	133
28.	Distribution of plankton in the North Atlantic in the summer . . . . *to face* p.	152
29.	Distribution of plankton in the North Sea in January, 1897 *to face* p.	156
30.	Variation of plankton and silicic acid in Kiel Bay . . .	237
31.	Forms of bacteria	256

PART I.
INTRODUCTORY.

CHAPTER I.

THE EXPLORATION OF THE SEA.

WHEN, in December 1872, the *Challenger* left Portsmouth on her memorable voyage methods of oceanographical discovery were very imperfectly developed and what was then known of the physics and biology of the deeper seas of the earth was neither very exhaustive nor exact. There had indeed been many voyages of discovery in the course of which researches other than purely geographical ones had been attempted, but it was only a few years previous to 1872 that what we now recognise as oceanographical investigation had been contemplated as an end in itself. Such a colossal task as a systematic survey of the oceans of the globe might usefully have been preceded by a preliminary study of the apparatus and methods of the earlier cruises, and by some extensive experiments applied to the perfection of these. This, however, was not done and the ship left England on her three years' voyage with an equipment containing little that was novel. But the expedition was planned out spaciously, the ship was powerful and commodious and her naval and civilian staff were efficient in the highest degree. The apparatus was thus tested in every possible way and an invaluable foundation was laid for later improvement.

About a dozen oceanographical cruises of first rate importance have been made since then and almost every detail of the *Challenger* equipment has undergone refinement and improvement.

J. F.

A perusal of Chapters II and III of the Narrative of the Voyage, and a comparison of these with an account of the latest voyage of discovery will reveal many improvements in details of apparatus and general methods. Nevertheless there is little essential change. The *Challenger* voyage was so prodigious an advance on anything that had previously been attempted that a corresponding amount of progress will be difficult to attain. If we except the modern quantitative method of attack on problems of marine biology there is little in the aims and methods of any of the latter-day expeditions that is not indicated in the instructions issued by the Admiralty to the *Challenger* staff. The ship proceeded on her three years' voyage without haste: she sounded, trawled, dredged and tow-netted, and when she returned a number of English and foreign specialists leisurely investigated the collections made, and collated and studied the data obtained. All this remains except the leisure: the modern expeditions sound, trawl, dredge and tow-net, just as the *Challenger* did, but with improved gear. To complete the parallel the famous *Challenger Report* still remains as the model on which all subsequent ones have been made.

The *Challenger* methods and apparatus have now acquired immortality in the cheaper text-books and I hope to acquire merit by refraining from their further description. A glance over the publications of the expeditions of the last thirty years reveals a host of improved forms of oceanographical apparatus. Every one of these cruises has added a new modification of some apparatus and has tacked on to it the name of some member of its staff, so that a catalogue of oceanographical gear would also be a list of the names of those who have taken part in this kind of discovery. For a description of all this apparatus encyclopediac limits would be necessary. How then is a conscientious author to indicate the progress made in this field? This chapter is, however, only intended to serve as an introduction to a more special aspect of marine study and therefore a somewhat cursory survey of the apparatus of oceanography will suffice.

The fundamental gear of the science are the sounding machine, the thermometer, the salinometer or its equivalent, the trawl and dredge, and the tow-net. This order is a logical one, since biological enquiry, so far as it is not a mere catalogue of the names, characteristics, and distribution of organisms, is to be based on the results of physical investigation.

The sounding machines. The names of these are legion. But when the industrious reader wades through the descriptions of those of Brooke, Baillie, Belknap, le Blanc, Massey, Sigsbee, Tanner and Thomson (to take only a selection in alphabetical order) he will find it convenient to consider only the latest—the Lucas automatic sounding machine—for this possesses merits which render it unnecessary (at least for our present purpose) to consider the others. In it are the characteristics of the best type of English machinery—elegance, simplicity, compactness and efficiency. Its evolution is the result of commercial enterprise. So long as a knowledge of the depths of the sea was a purely scientific study the ability and expense expended on effort to obtain it were restricted within obvious limits. But when the acquisition of this kind of knowledge became an "economic problem" invention galloped. When it became necessary to obtain an exact knowledge of the contour of the bed of the sea for the purpose of laying telegraph cables, the modern sounding machine came into existence. When further it became a matter of some considerable expense to stop an ocean liner on her voyage in order to obtain a sounding, the "depth-indicator" was soon evolved[1].

A "deep-sea" sounding may be regarded as one taken in water of over 200 metres in depth. To make this rapidly and exactly, and at the same time to obtain a sample of the deposits on the sea floor, is essential. Of necessity a very long sounding line is required, and this must be carried down to the bottom very quickly so that the line remains vertical. To do this a heavy sounding lead is attached to the end of the line, and obviously the latter must be strong enough to carry the weight. Yet the line must not be too heavy for it would otherwise possess in itself a very considerable mass. Now a thin line may be employed which will carry the lead when the latter is descending,

[1] In the depth-indicator a glass tube, open at one end, is lowered to the bottom. Water rises into the tube compressing the air. The depth is estimated by the application of Boyle's Law.

but when the line has to be hauled again its weight, plus that of the sinker, may conceivably be great enough to break the line. A further necessity of the line is that it should not offer much resistance to the water on descending or hauling. Again, the sounding machine must be so constructed as to stop automatically when the bottom is reached, and there must be a means of recording the length of line run out—that is the depth of the sea in the place sounded.

In the Lucas, as in all other modern sounding machines, the line is a pianoforte wire which is usually less than one millimetre in diameter. This is polished so as to offer little resistance to the

Fig. 1. The smaller Lucas sounding machine. By kind permission of the Telegraph Construction and Maintenance Company.

water in descending or hauling. The line is wound on a reel from which it runs out over a wheel which has a counter recording the number of revolutions, and therefore the number of fathoms, since the diameter of the wheel is known. This arrangement is shewn in Fig. 1, which represents the smaller Lucas machine.

The recording wheel is attached to a movable arm which is connected with the frame of the apparatus by two coiled steel springs. When the weight of the sinker is on the wheel the arm is extended stretching the springs. This releases a brake which acts on the reel containing the line, so the latter runs out freely. But immediately the sinker touches the sea bottom the weight is taken off the arm and the springs pull the latter back while the forward movement of the base of the arm tightens a brake which lies close against the rim of the reel, whereupon the latter is stopped. The counter is then read and the depth of the sea is obtained.

The old-fashioned sounding lead of the mariner is a lead cone of about 7 to 30 pounds in weight. Remember that a very heavy weight is necessary to carry the line down rapidly to the bottom and prevent the latter from being thrown into coils or kinking. But when we add to the weight of the sinker that of the sounding line there is a very considerable strain on the latter and this is enough to break any line of manageable dimensions when a deep sea sounding is being taken. It is therefore necessary to devise some means whereby the weight, when it has served its purpose of carrying the line to the bottom, may be released and fall off, when the latter can then safely be reeled up again. All sounding leads are constructed with this object and all are modifications of that first devised by "Passed Midshipman John M. Brooke of the U. S. Navy." I describe here the latest form, the "Sondeur à Clef" of the Prince of Monaco.

This is shewn in Fig. 2. It consists of a gun-metal tube furnished with a stop-cock at its lower end. The handle of the latter is a lever which, when the cock is closed, fits into a depression cut in the side of the tube. When the cock is open this lever stands out at right angles as is shewn in Fig. 2 on the left. The opening of the stop-cock is of the same bore as the tube. At the upper end of the latter is a plunger which can move up and down for a short distance—the latter being regulated by a slot cut in the sides of the tube in which work pins attached to the plunger. The sounding line is attached to the plunger and on the sides of the latter are two notches. When this plunger falls down within the tube the notches sink below the rounded upper end of the latter. The sounding tube itself is comparatively light and the weights (which may be 20 lbs. each) are slung on to the notches in the plunger. It is these weights which carry down the sounding line. Their total mass varies with the depth of water.

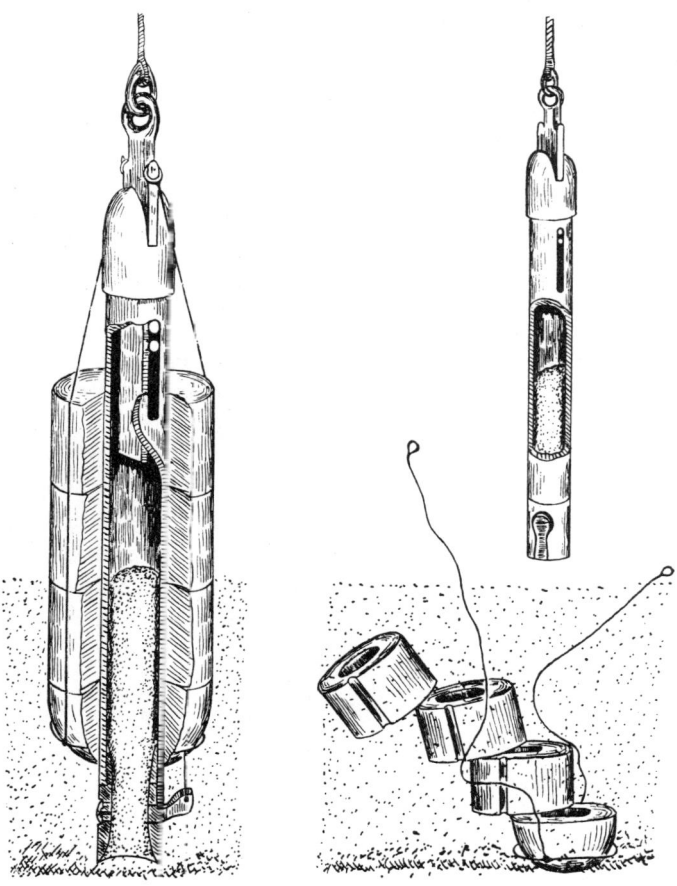

Fig. 2. The stop-cock sounding lead of the Prince of Monaco. There are several forms and that figured is described in the *Campagnes Scientifiques*, pp. 14 and 15. On the left the sounder is shewn descending, and just touching the bottom. On the right the sounder is shewn ascending. The weights have been released; the tube is full of the bottom deposits and the stop-cock is closed.

When the apparatus is to be used one or more weights—cast iron rings—are slipped over the outer tube, and the whole thing being suspended these rings are secured to the lead by means of a sling which passes over each of the notches in the plunger. It will easily be seen that when the lead touches the bottom the tension of the line decreases and the plunger drops within the tube. When the notch in which the sling is placed falls below the opening of the tube it releases the weights, which then drop off and are lost. But in falling off they push down the lever of the stop-cock which then closes. The tube has however been pushed down into the sea floor and is filled with a sample of the deposit there, and the closing of the cock prevents this from falling out when the line is hauled up again.

The water sample bottle and the thermometer. A simple sounding thus gives us a knowledge of the depth of the sea and also a sample of the deposits lying on the sea floor. A "hydrographic sounding" gives us a sample of the sea water at the bottom or at any required depth, and in addition the temperature of the water *in situ*. The water sample is to be used for the determination of the density, salinity, gas contents, or any other property or constituent of the water. There are a host of water bottles (Buchanan, Ekman, Jacobsen, Kidder-Flint, Meyer, Mill, Regnard, Sigsbee, Wille, and probably others). It will serve the reader best if I describe only the water bottle which is used by the vessels of the International Fisheries Investigations, and which was devised by Nansen and Pettersson.

This instrument is represented diagrammatically in Fig. 4. It consists essentially of two brass rods connected together at either end, and between which slides an open cylinder, which is the actual bottle. Fixed between the lower ends of these rods is a stopper which can accurately close the lower end of the cylinder; and sliding freely on the rods above the cylinder is another similar stopper which can close the upper end of the bottle. The wall of the latter consists of four concentric cylinders of metal and ebonite alternately: each stopper consists of a metal cap carrying three rubber discs and it is the latter which fit into the open ends of the bottle. The upper stopper is perforated to receive a thermometer, and the lower one is provided with a stop-cock. When the bottle is about to be lowered into the sea the parts are

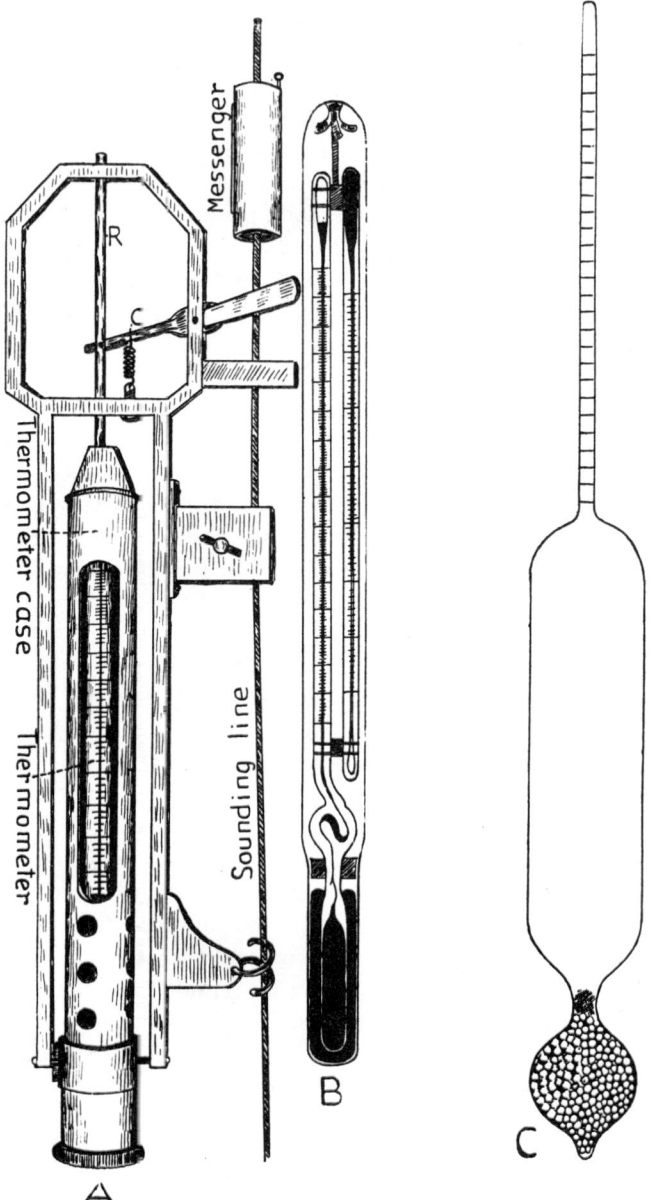

Fig. 3. A. The Negretti and Zambra deep-sea reversing thermometer. The messenger is lowered down the line, strikes the catch C which then lifts the rod R from a depression in the thermometer case. The latter then turns over or reverses, breaking the mercury thread in the thermometer. B. The most recent deep-sea reversing thermometer (Richter's pattern). C. A Kiel hydrometer

fixed in the position shewn by means of a spring catch not shewn in the diagram. The whole is attached to a steel wire rope and when it has been lowered to the desired depth a "messenger" is allowed to slide down the line until it strikes against the catch, when the cylinder slides down the rods until it is closed by the

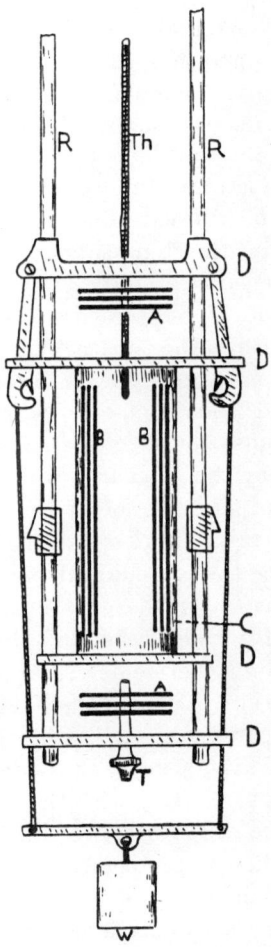

Fig. 4. The Nansen-Pettersson water bottle. Some parts are omitted. A, A are the stoppers; B the concentric walls of the bottle; C the outer case of the latter; D, D, D, D the ends of the stoppers and the bottle, these parts slide on the brass rods R; T is the stop-cock through which the water sample is withdrawn; Th is the thermometer, and W the weight which carried down the apparatus.

lower stopper: at the same time the upper stopper also slides down the rods and closes the upper end of the bottle. Two spring catches shewn at the sides then make fast the whole apparatus. A sample of water at the depth to which the bottle is lowered thus fills the cylinder[1].

The bottle therefore collects a sample of water and isolates this in a central chamber surrounded by three concentric shells of water at the same temperature as that of the sample. These shells of water insulate the sample, preventing the conduction of heat from without, if, as is usually the case in a deep water sounding, the water from the deep has a temperature which is above that of the water strata through which the bottle has to be hauled in coming to the surface. The same sounding therefore enables us to ascertain the temperature of the water while collecting a sample for further study.

The determination of the temperature of the sea is obviously a simple matter when it is the surface layers which are being investigated. All that is necessary is to lower a bucket over the ship's side, fill it and then immerse a thermometer in this. Obviously the sample must be collected from such a situation that it is not contaminated by the discharge from the engine-room or from any exit-pipe from the interior of the vessel. But when a temperature has to be taken from a water layer which is at some distance from the surface the procedure is much more complicated. In the time of the *Challenger* the only means of obtaining this value was by the use of a "slow-reaction" thermometer, by using a maximum and minimum thermometer, or by using a reversing instrument. There were indeed other ingenious methods of obtaining deep-sea temperatures, such as by the electrical thermometer of W. Siemens. This depended on the fact that the electrical resistance of water varies inversely with the temperature. If then we balance the resistance of the water *in situ* with that of a sample of water on board at a known temperature by means of a Wheatstone bridge the problem is solved. But this method demands delicate apparatus not easily managed in a ship tossing in a seaway, and as a practicable means of obtaining deep-sea temperatures it has now been abandoned. The slow-reaction

[1] The water-bottle is described in detail in *Publication de Circonstance* of the International Council for the Exploration of the Sea, No. 21.

thermometer was one in which a long time was required for the instrument to indicate the temperature of the medium in which it was placed. Therefore in hauling it to the surface the temperature did not greatly change and the time increments of change could be calculated. But a practical objection was the time required to make an observation. So this method also has been abandoned. Finally we have the reversing thermometers, and these are still used. The instrument of this class originally made by Negretti and Zambra, Fig. 3, is too well known to require description. We have a thermometer tube which is constructed in much the same manner in which a clinical thermometer tube is made, that is the bore of the tube immediately above the bulb is bent twice in an S-shaped curve and is greatly contracted at this point. In a clinical thermometer the thread of mercury breaks when the instrument is taken out of the place the temperature of which is desired, and when the metal contracts with the reduction of temperature. The mercury in the capillary tube therefore remains where it was. In a deep-sea thermometer in which this principle is adopted the case carrying the tube is reversed when the desired depth is attained, this being done usually by a small propeller in the outer frame of the instrument which revolves immediately the direction of movement of the machine is changed on being hauled, and then releases a catch which allows the tube to reverse. The mercury then falls to the other end of the tube and as that latter is graduated from this end the temperature is recorded. In the Nansen-Pettersson water bottle a reversing thermometer is usually attached to the frame of the instrument and is reversed when the messenger springs a catch at the same time as closing the bottle. But since the thermometer enclosed in the central chamber of the bottle takes the temperature of the water *in situ*, and since this sample of water does not change greatly in temperature because it is surrounded by a practically non-conducting material, the indications of the reversing instrument are only used as a check on the other. The sample of water obtained by the Nansen-Pettersson bottle does actually change in temperature as the instrument is being hauled through water of a temperature markedly different from that to which it has been lowered, but the change is very slight and can be allowed for.

The water bottle which I have just described is a triumph of mechanical skill and it is greatly to the credit of the International Fishery Organisation that it has placed at the service of oceanography an apparatus of such efficiency.

It is then possible to obtain (1) a sounding at any depth of the sea, (2) a sample of the deposits at the bottom, (3) a sample of the water at the bottom, or at any desired depth, and (4) a determination of the temperature of the water at the bottom or at any depth. Knowledge of the density, chemical composition, and salinity of the water samples studied is also required for the statement of the physical characters of the sea in the region investigated.

Formerly the important values, salinity, specific gravity, and density were obtained by means of the salinometer. The salinometer is really a hydrometer adapted to the study of sea water. The Kiel "araometers," Fig. 3, are hydrometers constructed to give the specific gravity of sea water. Since they are used in water of very different degrees of salinity a set of six is required to cover the range of variation. The instrument is a double glass bulb. The upper bulb is long and capacious and contains air only; the lower bulb is weighted with lead shot. At the upper end of the upper bulb is a long stem in which is a graduated scale. Each instrument is constructed to float in water of a certain mean density in which it is immersed up to the middle of the scale, the whole thing at the same time floating freely in the liquid. If the density is slightly greater the araometer rises, if it is lighter it sinks. The readings given by the araometer are values representing the specific gravity of the water at the temperature at which the observation is made.

The specific gravity of a sample of sea water is the number representing its weight as compared with that of an equal volume of pure water at the same temperature. The latter is usually called 1·000 so that the specific gravity of a sample of sea water may be some such number as 1·025. The density is the weight in grams of one c.c. of water at the temperature *in situ* ($t°$) compared with that of one c.c. of pure water at 4° C. It is usually expressed as

$$S\frac{t°}{4°}.$$

The salinity is the total weight in grams of the solid matter dissolved in 1000 grams of water.

Now sea water is a solution of a great number of different salts of which common salt is the most abundant. But it has been found that the ratio of these different substances to each other is always very nearly the same even if we are dealing with sea water of very variable degrees of saltness. This is so because the sea is salter or fresher by reason of the removal of pure water from it by evaporation, or by the addition of water to it from the land, not quite pure but practically so because it becomes mixed with such an immense quantity of salt water that the ordinary constituents dissolved in the fresh water are immensely diluted. So it has been possible to draw up tables which enable us to convert one of the values referred to, say salinity, into any other, say density. Only one determination is then required in order to define the character of the water sample, and this is now always the total contents of the water in dissolved halogens. Not only chlorine, but also bromine and iodine are present in solution. The total halogens are then estimated by means of titration by nitrate of silver, a method which is susceptible of a high degree of accuracy, and from this value the other values—specific gravity *in situ*, and density—are calculated.

The practical procedure adopted in the International Fishery Investigations is as follows: the sample of water is obtained by means of the Nansen-Pettersson bottle and at the same time the temperature of the water *in situ* is taken. The water sample is taken from the bottle and stored in a clean glass tube or bottle and sealed. These samples are then sent ashore to be worked up. Standard solutions of silver nitrate having previously been prepared the titration is carried out. This is done in a special burette consisting of a wide glass bulb containing the greater part of the silver solution required for the titration. The narrow stem of the burette is graduated so that the estimation of very small quantities of the solution delivered is easy. The burette gives direct readings—that is, it indicates the amount of total halogen represented as chlorine in the water sample examined in parts per 1000. Standard sea water is prepared by the Central Laboratory at Christiania and is supplied to the workers in the various countries, and this water being investigated as to its chemical composition with the utmost care, is used for the standardisation of the silver solutions employed in the analyses. It will be seen then that most scrupulous care is taken that the values employed in the discussion of the oceanographical problems of the North-Atlantic Ocean are determined with very great accuracy: the temperature is in fact determined to 1/10th of a degree C., and the salinity with an accuracy of 0·05 in 1000 parts, the density being thus capable of calculation to within 0·00004. These observations—depth, temperature and salinity—are at the present time being made simultaneously over the whole of the North European seas. England investigates the Channel, Scotland the Northern part of the North Sea and the Faeroe Channel, Ireland the western part of the Irish Sea and the Atlantic Ocean to the West of Ireland, Germany parts of the North Sea and the Baltic, Belgium part of the North Sea, Denmark the Cattegat and Belt Seas, Holland part of the North Sea, Norway the Norwegian Sea, Russia the White Sea and part of the Baltic, Sweden the Skagerak and part of the Baltic, and Finland the Gulf of Bothnia. The reader should refer to the chart (Fig. 5) of the International Fishery Investigations area. The various countries make cruises once a quarter or oftener and the results of the analyses of the water samples collected on these cruises as well as the temperatures observed, and the principal plankton organisms present in the sea at the time of the cruises, are then communicated to the International Fishery Council and are published and discussed by the latter[1].

These determinations of the physical and chemical nature of the waters of the seas by no means exhaust the amount of information that a study of the properties of that liquid can afford. A knowledge of the amounts of dissolved gases such as oxygen, carbonic acid, sulphuretted hydrogen, etc., is often required. So also with the dissolved nitrates and nitrites and ammonia salts; the sulphates, phosphates and carbonates, the dissolved lime and silicic acid. I am sure that the reader will excuse me from a recital of the means of determination of these various constituents of sea water. I need only observe here that all are required in the discussion of the problems connected with the biology of

[1] In the *Bulletins des Resultats*.

marine organisms, and that means are being, or have been, devised for their estimation; by no possibility an easy matter since all are present in small quantity and some are in very minute proportions indeed. Then I need only refer to other physical investigations relating to the problems of the sea, such as the relation of the viscosity of sea water to the movements of minute organisms, or to investigation of the physical nature of the deposits forming the sea floor in relation to the thermal reactions of the former.

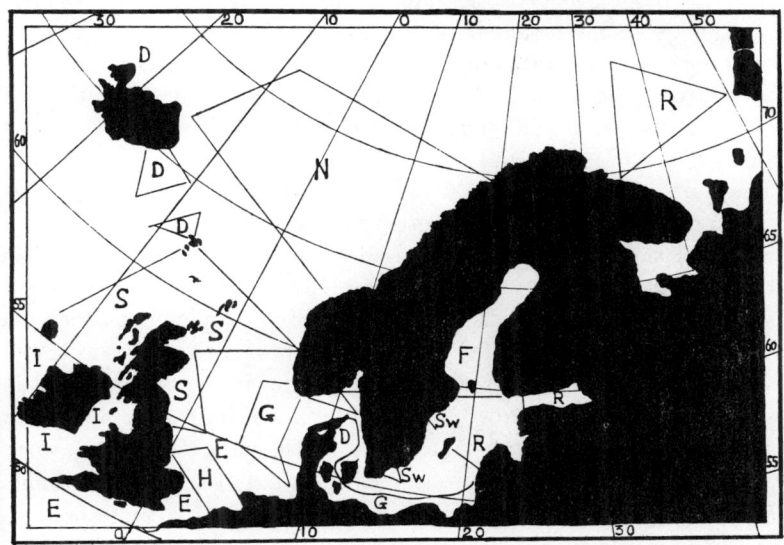

Fig. 5. Sketch Chart of the International Fishery Investigation area shewing the division of the observations between the various countries. *D*, Denmark; *E*, England; *F*, Finland; *G*, Germany; *H*, Holland; *I*, Ireland; *N*, Norway; *R*, Russia; *S*, Scotland; *Sw.*, Sweden.

Salinity and temperature, these are the properties by a consideration of which it has been possible to trace the movements of large masses of sea water in the area of the North European seas. But direct estimations of the direction and rate of these movements are also possible and have indeed been made. Surface and bottom floats have been employed in order to ascertain the flow, the former by many investigators but notably on a large scale by the Prince of Monaco[1]: and the latter first successfully

[1] See the *Campagnes Scientifiques*.

by Bidder in the North Sea[1]. Then we have several most ingenious pieces of apparatus devised in order to estimate the rate of flow of slowly moving surface and bottom currents. The elucidation of the mode of action of these apparatus I leave to the reader[2].

We have considered these methods of physical observation since they are fundamental to any attempt to discuss what we may term the metabolism of the sea. We have now to consider the methods by means of which the biology of the sea has been investigated. I except the laboratory methods of study of the physiology and development of marine organisms and deal here only with the apparatus of collection. Marine creatures have the most diverse habits of life: some live in the mud at the bottom of the sea, others on the surface of the deposits there, others at the surface of the water, and others again at all intermediate levels. They are of great diversity of size, from that of a micrococcus (1—5 thousandths of a millimetre) to that of a large fish or marine mammal such as a whale. Therefore our collecting apparatus has to be adapted to the mode of life and size of the creatures which it is desired to obtain. Ordinary fishing implements may be classified into bottom fishing instruments such as the trawl, dredge, line or set-net; and into pelagic implements like the seine-net, drift-net or tow-net. The former fish at the sea bottom and catch the organisms which live there, while the latter work at the surface or at any intermediate level and catch the animals or plants which live a pelagic life swimming or drifting about in the body of the sea.

The dredge and trawl. The dredge, Fig. 6, is the older instrument. It consists of a rectangular iron frame which is from

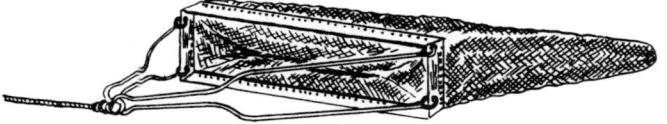

Fig. 6. The naturalist's dredge.

[1] *Rapports et Procès-Verbaux; Conseil Perm. Internat. Exploration de la Mer,* Vol. VI., 1906.

[2] F. Nansen, *Publications de Circonstance, Conseil Perm. Internat. Exploration de la Mer,* No. 34, 1906, Copenhagen. The action of these machines is very interesting and well worth study by the reader.

two to five feet in length and about nine inches to two feet in breadth. To this frame is attached a bag of twine netting, the mesh of which is always fine, but varies of course with the purpose for which the dredge is intended. From either end of the frame an arm is fastened which passes forward and to which the dredge rope is attached. Only one of the eyes at the ends of the arms is attached to the dredge rope; the other is attached to the eye which is shackled to the rope, by a lashing of rope yarn, the object of this device being to free the dredge in case it may become fast on some obstacle on the sea bottom. If this occurs the direction of strain on the dredge is altered and the latter usually becomes free. Sometimes a bar is attached to the end of the bag and to this are fastened a number of swabs or tangles of yarn; to these many delicate objects become fast which otherwise might be damaged by the accumulation of débris in the bag.

Even a small trawl is a more efficient instrument than the dredge, although it may be a less suitable instrument for some purposes than the latter. In Fig. 7 I give the form and dimensions of a trawl used for scientific purposes. The trawl

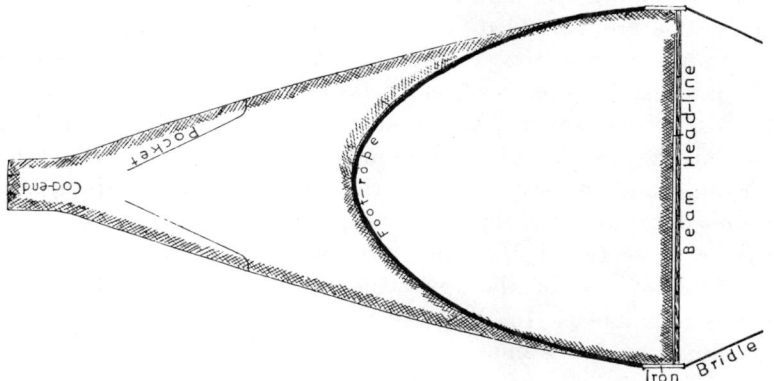

Fig. 7. Diagram of a small beam trawl used for scientific investigation.

consists of two "irons," stirrup-shaped contrivances which are joined together by a beam of wood. The length of this beam may be about 50 feet in the largest commercial trawls carried by fishing smacks. The flat side of the iron runs on the sea bottom and the beam is about two feet from the ground. Behind the beam and fastened to it and the irons is the "headline," and connecting the lower parts of the irons is the "foot-rope," a rope which is about three inches in diameter, which is often weighted with iron chain, and which drags on the sea bottom, forming a wide bight behind the beam. To the foot-rope and headline is "bent" the trawl net. This is a long conical bag of strong netting, the mesh of which varies with the size and nature of the animals which the trawl is designed to catch. The length of the net varies but it is always one and a half times or more than the length of the beam. About half-way between the foot-rope and the end, or "cod end" of the net, are two folds of netting attached to the back and belly of the trawl net and inclined backwards towards the middle line of the latter but not meeting. These are the "pockets," and they form a funnel-shaped trap through which the animals caught by the net pass into the tail and through which they find difficulty in passing out again on account of the eddies caused within the net by the water which does not pass through the meshes. The cod end is open but is laced by a rope when the net is put down. From each of the irons a strong rope, the "bridle," passes forward and the ends of the bridles are "shackled" on to the trawl warp or rope.

In the largest trawls used by fishing steamers the beam is discarded and the ends of the net are kept apart by two "otter boards," which are heavy wooden framed boards shod with iron and about the size of an ordinary house door. To these the headline and foot-rope are attached. The otter boards are so fastened to the two separate trawl warps, by means of which the trawl is towed, that they are inclined at an angle to the direction of tow of the apparatus, and thus the mouth of the net is kept open. In the largest otter trawls the length of the headline may be as much as 100 feet, the foot-rope being half as much again.

For scientific fishing, that is fishing with a scientific object a "shrimp trawl" is usually more suitable than the ordinary wide meshed fish trawl. In the latter the meshes are usually about one to one and a half inches from knot to knot, while in the shrimp net the distance is only about half an inch. Otherwise the form of the two instruments is the same. A useful modification of the shrimp trawl is the "shank net," which is essentially a trawl, the mouth of which is kept open by a frame of wood about ten feet long by one and a half to two feet wide and one side of which drags on the ground.

In a small steamer or a sailing vessel the trawl is towed by a rope of hemp or steel wires which passes over the rail through towing chocks and is coiled loosely round the bollards. The trawl rope is attached to the bollards by means of a "stopper." This takes the strain of the tow and should the trawl catch on any obstacle on the sea bottom and be "brought up" the increased strain is thrown on the stopper, which then parts, warning being then given to stop the vessel and get the net loose. In this case the slack of the warp is taken in and the vessel then sails round the obstacle, when the trawl is usually detached.

In Fig. 9 I give a diagram of the arrangement of the trawl, &c. on the deck of the ship when it is in readiness for "shooting."

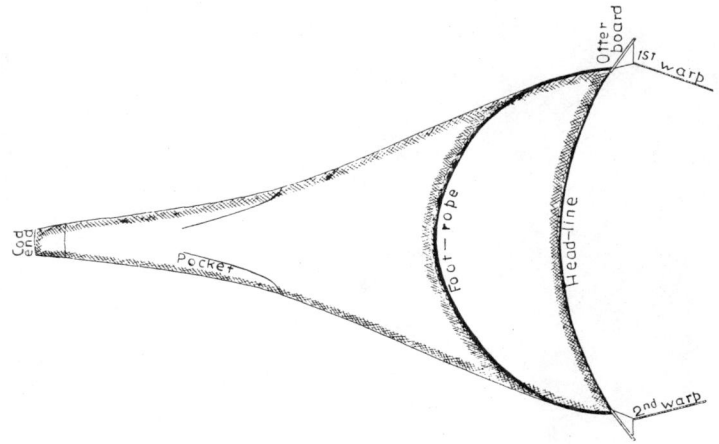

Fig. 8. Diagram of an otter trawl.

The beam lies along the starboard rail, the after iron projecting over the counter and the forward iron just inside the rail. The latter is lashed to a stanchion and the former to a small davit over which the rope is passed when the net is being hauled. The bridles are loosely coiled up on deck out of the way of anything on which they might catch. The trawl rope is coiled round the starboard bollards and is passed aft through the towing chocks so that the shackle to which the bridles are attached is without the rail. A long rope, the "head-rope" has been rigged so that

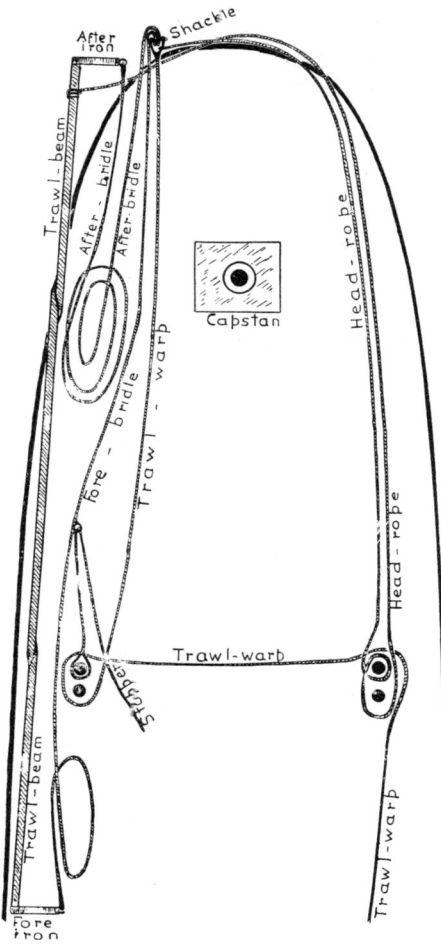

Fig. 9. Arrangement of the trawl on the deck of the Lancashire surveying fishery steamer *John Fell*. The whole arrangement is one adapted for a small vessel such as a steam yacht, and involves little alteration of the deck fittings. Otter trawls are usually employed on yachts, but since the small beam trawl represented can be easily taken to pieces and stowed away out of sight, its use is to be preferred.

one end of this is attached to the beam near the after iron, while the other end is attached to the shackle: the bight of this rope is brought round to the port bollards and is coiled round these.

About 12 feet from the forward iron a thimble has been worked on to the forward bridle and through this the end of the stopper is passed, and this rope is then made fast to the starboard bollards. "Shooting" is then carried out as follows: (1) the slack of the net is thrown over the side; (2) the forward iron is thrown into the sea and is brought up by the stopper when about 12 feet of the after bridle has gone overboard; (3) the after iron is now also thrown over the side and at the same time the stopper is let go, the net now drifts out astern and is brought up by the head-rope; (4) this too is let go and the trawl now is "squared up" as it drifts out astern. The trawl warp is then slowly paid out until the officer in charge of the operation is satisfied that sufficient is out to allow the net to drag nicely on the bottom. The stopper is then put on and the ship steams at a speed which is adjusted to the nature of the fishing. The whole operation is delicate and requires much skill and experience, not only that the trawl may be successfully shot—that it may not for instance capsize and fish upside down, in which case the whole operations are futile—but also that it may not result in an accident to some one of the crew of the trawl, by no means an improbable contingency, should the shooting be carried out carelessly[1].

When the trawl is to be hauled the ship is stopped and the stopper is taken off. The trawl rope is then passed round the capstan, the latter is started and the rope is hauled in. When the bridles have passed round the capstan these are watched until the thimble on the forward one has come to the surface; a block is then made fast to this and the forward iron is hauled by hand, while the head-rope, which is attached to the beam near the after iron, is also hauled by the capstan. The beam comes to the surface, is hauled to above the level of the rail, is lowered over this and is made fast. The slack of the net is then hauled in by hand if a light catch has been made, or is lifted by a tackle if much material is contained in the net. Finally the latter is brought on deck, the cod-string is unlaced and the catch is shaken out.

---

[1] I am indebted to my friend Captain Wignall for this account of the practical working of the trawl net. The diagram on page 20 has also been prepared by him.

Deep-sea dredging or trawling presents difficulties of a very real kind. First of all an immense length of trawl rope is necessary to lower the net down to the depths at which it is proposed to fish. To obtain a strong, yet light rope is in itself a problem. Formerly hempen rope was used, thus the *Challenger* used three-inch rope of this nature. Nowadays steel wire rope is exclusively used, the compactness, strength and ease of stowing of this being undoubted. Nevertheless the *Challenger* dredged in 3875, and trawled in 2650 fathoms. Then in deep-sea trawling or dredging a large and heavy ship is of necessity used, and the momentum of this presents other difficulties. For a small ship will give and take to the variable tension of the trawl rope, but in a large vessel a sudden movement of the latter, such as is caused by a sea striking her, may easily put such a tension on the trawl warp that the latter may part, with the loss of part of it as well as the fishing instrument. This is avoided by the use of the apparatus known as the "Accumulator" or Dynamometer. In the *Challenger* this was constructed of two strong discs connected together by forty cords of rubber each one inch in diameter. To each disc a number of lanyards were attached; one series of these were connected with a block which was secured to the foreyard near the end of this. The other series of lanyards were connected with a block which carried the dredging rope. When the ship rose to a sea the rubber cords in the accumulator were extended and some of the increased tension was taken off the trawl rope. Latterly strong coiled steel springs have been used in place of the rubber cords, and in some cases rubber pads acted as the material used to take the extra strain. But the management of the trawl rope and fishing gear in deep-sea work involves problems which are quite outside the scope of this work, and for some account of the methods involved I may refer the reader to the description of the fishing gear used in the *Siboga* Expedition[1], or to the admirable work on deep-sea trawling in the description of the United States research vessel *Albatross*[2]. In deep-sea trawling the form of

[1] See Reports of this Expedition. Livr. IV. gives an account of the equipment of the ship, 1902.
[2] See Tanner, *Bull. U.S. Fish. Commission*, Vol. XVI. (for 1896), 1897. This paper is full of interest.

the trawl is modified, and instead of the instrument described in this chapter a trawl frame is employed which fishes equally well no matter on which side it reaches the bottom, an important consideration, since it is exceedingly difficult so to manipulate a trawl in deep water that it reaches the bottom on one particular side.

Fig. 10. The so-called "Agassiz trawl."

Success in trawling, even in shallow water, depends on a variety of circumstances, such as the length of line used, the exact construction and trim of the apparatus, the speed of the vessel, &c. Even with the same apparatus fishermen of different degrees of experience and skill obtain very variable results on ground which might reasonably be expected to give uniform results, a consideration which does not receive due attention in discussions of trawling results applied to the elucidation of the abundance of fishes or other organisms on a sea area from time to time.

In fishing with a trawl net we catch whatever is upon the sea floor, except of course such animals which are able to swim out of the trawl when once they have been caught in it. Then the net usually catches a larger or smaller mass of bottom invertebrates, weed, sand and mud, stones, &c. When the catch is unusually large and the net is hauled for a long time many of the more delicate organisms are injured or killed by the pressure of the other materials in the catch, although this destructive action of the trawl is far less than one might imagine on a first consideration of its mode of action. When it is desired to obtain organisms which the trawl might injure fishing baskets are sometimes employed, a method of scientific fishing which was first systematically adopted by the Prince of Monaco[1], though it is really only a modification of the primitive lobster basket or creel. The fishing basket or "Nasse" is a frame of wood or iron which may have various forms: that employed by the Prince of Monaco was a short triangular prism. It is covered with netting and in this are openings of various sizes, or there may be inserted baskets of different sizes of

[1] Richard, *Les Campagnes Scientifiques*, Monaco, 1900.

mesh. The openings are guarded by pockets or valves which admit the unlucky animal which is impelled to explore the apparatus, but refuses it exit. The basket may be baited in the manner of a lobster pot: it may even have an electric lamp within it to act as a lure to the denizens of the deep. The basket is

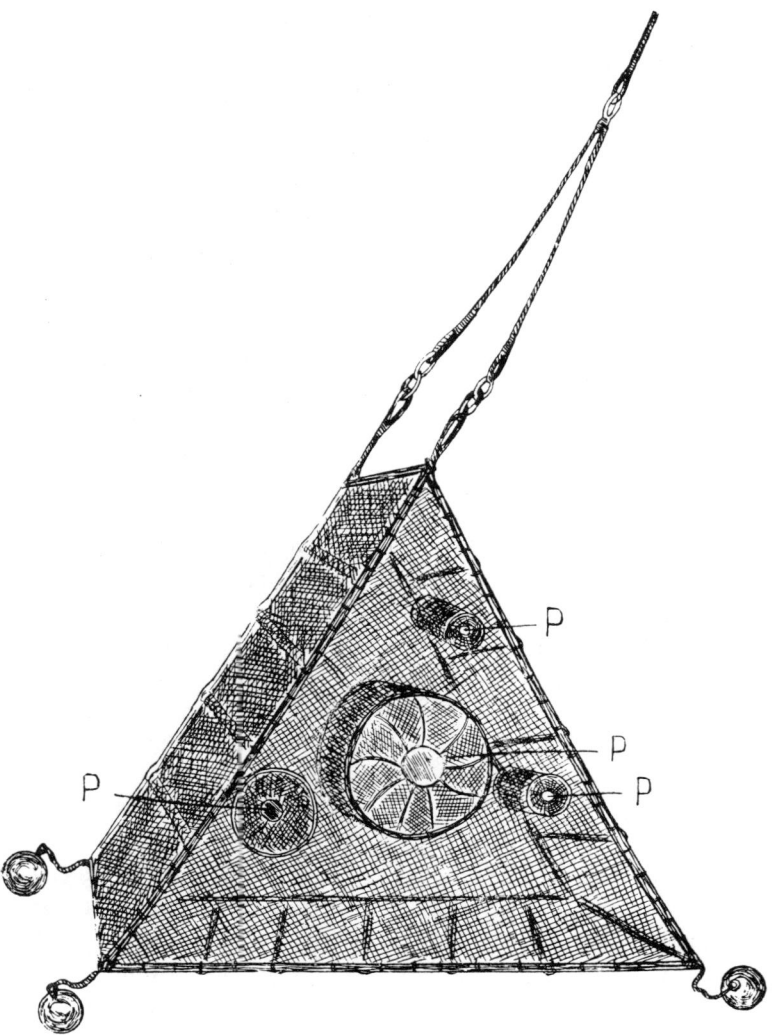

Fig. 11. Sea bottom fishing basket (from the *Campagnes Scientifiques*); P = pocket.

weighted and lowered to the bottom where it remains for some time with a buoy attached to the end of the rope.

Stake nets, trammel nets, lines and the various forms of fishing weirs or baulks belong rather to the province of the practical fisherman than to that of the collector. But these instruments have made their contribution to the accumulations of our museums and to the zoological lists, and therefore deserve some notice. The stake net is a vertical wall of netting about three feet in height and several hundred yards in length. It usually forms several angles and at these traps may be placed. The net is supported on a row of wooden stakes. It is placed in such a situation that it is bared by the receding tide when it is "fished."

The trammel net consists of three nets hanging parallel with one another. In the middle there is a net of small mesh, and on either side of this is a net which has meshes of larger dimensions. The three nets are placed close together on the same top and bottom lines. It is much higher than the stake net. The lower rope being weighted by sinkers and the top one buoyed by corks, the apparatus is sunk in the sea. Fishes striking against the inner net force this through the wider meshes of the outer ones. They are thus enmeshed, or "trammelled."

Lines are set on the sea bottom, being secured by sinkers and buoyed at intervals. Along the length of the line are the "snoods," short pieces of cord to which the baited hooks are attached. The "long lines" of the fishermen may be as much as six or seven miles in length.

Fishing weirs and baulks are arrangements of basket or wicker work, furnished with pockets or traps, and set across a tidal channel or gutter. The action of these apparatus is very simple. Fishes or other animals are carried against them by the tide and are caught in the meshes or pockets.

Pelagic fishing apparatus. The forms of fishing apparatus which we have been considering catch animals which reside at or near the sea bottom but are not adopted for the capture of those which lead a pelagic life—that swim about at the surface or at intermediate levels. It is true that the trawl may be used for the capture of pelagic fishes; thus quite recently steam trawlers have used their otter trawls for the capture of herring, by so manipulating them that the nets sweep through the water before descending to the bottom, or fish a considerable part of the sea while being hauled. But as a rule quite different kinds of fishing apparatus have to be employed for the capture of pelagic animals. The drift net and the seine employed in the Firth of Clyde for the capture of herring are commercial fishing instruments and are little used for scientific investigations. The drift net is simply a vertical wall of netting of great dimensions which is allowed to drift in the sea near the surface. Fishes striking against it are enmeshed, since the diameter of the meshes is less than the average diameter of the body of the fish. The seine net used in Loch Fyne is a large net which is "shot" round the shoal of herring, and is then hauled so that it forms a huge pocket in which the fish are contained.

The tow-nets. The tow-net of the naturalist, that is the traditional instrument, is a conical bag of some fine fabric, such as muslin or silk, or even fairly coarse canvas. This bag, which is about three or four feet in length, is attached to a ring of iron by means of a wide hem or strip of some stronger material. The end

Fig. 12. The ordinary surface tow-net.

of the net may simply be the termination of the cone of fabric, or there may be a bottle of metal into which the catch of the net is washed when it has been hauled. The net is towed by three bridles which are attached to equidistant parts of the ring, and which are then attached to the tow rope. It is towed at a low rate of speed. When the tow is completed the net is hauled, turned outside in and the catch, which consists of small, usually microscopic organisms, adhering to the walls of the net, is rinsed off into a bottle of water and these are then studied fresh or are preserved. With fine material, such as fine silk, the tow-net catches little and the catch is always small "fry." This is because of the restricted "draft" of water through the meshes of the net. With nets made of coarser material more may be caught but obviously the catch consists of larger organisms. The material of the tow-net is thus of necessity adapted to the nature of the organisms which it is desired to catch.

This is the instrument consecrated by long use, so that in some quarters it is a heresy to suggest that it is, for the purposes of modern investigations, rather an inefficient form of apparatus. During the last few years, mainly under the stimulus of the International Fishery Investigations, a number of new forms of fishing apparatus, modifications of the older tow-net, have been devised in Norway and Denmark and these have led to a great extension of our knowledge of pelagic organisms of various kinds.

The Heligoland young-fish trawl, to which, fortunately, the name of a zoologist is not prefixed, is a bag of canvas, or some other comparatively wide-meshed fabric, which is attached to a square frame of metal tubing. The opening of this frame is 72 × 72 cm., and the length of the net is 3 metres. To the under edge of the frame is the characteristic part of the apparatus, the

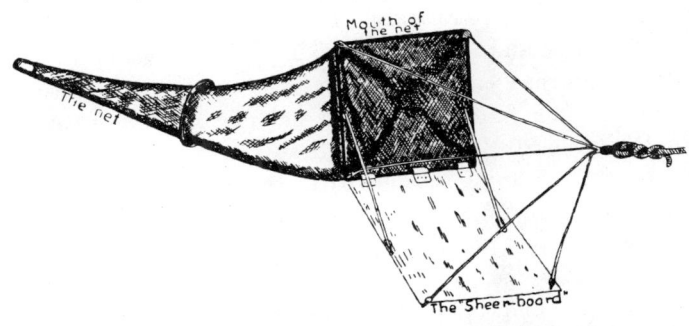

Fig. 13. The Heligoland "Scherbrutnetz." (See *Wiss. Meeresunt. Kiel Komm. Abth. Helgoland*, Bd. VI. Heft 1, 1904.)

"sheering board." This is a square piece of galvanised iron plate which is hinged to the frame and is of the same dimensions. It can be set at an angle to the frame, usually 125° inclining forwards and downwards, and is fixed in this position by wire stays. The whole is towed by six wire bridles bent to the usual rope. The inclined board gives the whole net a sheer downwards whereby it fishes in the deeper layers of the sea even near to the bottom, or by suitable management at any required depth. Its comparatively wide mesh and its easily controlled position make it an instrument of the greatest value for fishing for the larvae and young stages of fish, or indeed for many other organisms of larger size than those usually caught by the ordinary tow-net.

Then we have the various modifications of the otter trawl devised by the Heligoland, Norwegian and Danish naturalists. I leave the reader to puzzle out for himself the synonomy of these instruments. One is a long conical bag of hempen net which is bent on to an iron ring. The net is about $5\frac{1}{2}$ metres long, and the diameter of the ring about $2\frac{1}{2}$ metres. It is towed by three bridles and a steel rope, and is kept at the surface or at any depth by a float. A modification of this apparatus is a larger net conical in shape, with a rope ring to keep it open, to which are attached three otter boards from which three bridles pass to a steel warp. These three otter boards keep the mouth of the net open so that the opening is a triangular one of about 50 to 100 square metres. The length of the net is about 30 metres.

Finally we have the young fish trawl used by the Danes. This is a pelagic otter trawl differing from the parent instrument mainly in its size and in the fineness of the mesh, which is adapted to catch organisms of the size of very young fishes. All these instruments differ from the old tow-net (1) in their greater size, (2) in their wider mesh whereby they are enabled to catch the larger planktonic animals, and (3) in the ease whereby they can be made to fish at different levels in the sea. Some consideration of the literature of the International Fishery Investigations will shew that it is by means of these instruments that very important additions to our knowledge of the life histories of fishes have been made, additions that we should probably still be waiting for had we possessed only the old tow-nets as means of investigation.

All the fishing apparatus I have described are to be characterised as qualitative apparatus, adapted only for the purpose of ascertaining the presence and distribution of marine organisms.

Forms of fishing gear capable of giving quantitative results do exist but the consideration of these will occupy us further on. In the present chapter I have dealt with the exploration of the sea so far as this may be studied by the exact methods of physical and chemical science; and with the ordinary collecting methods of biology, which so far aim mainly at the discovery of the forms of life which are present in the sea, and with the broader facts of the distribution of these.

CHAPTER II.

THE OCEANOGRAPHY OF THE NORTH-WESTERN OCEAN[1].

The British Islands and the Continental Shelf. It is well known that the British Islands are situated on an elevated portion of the ocean bed which forms a continuous edging round the continent of Europe—this is the "Continental shelf" or "Continental slope." If we draw an imaginary line round the British Islands which connects all places on the sea bottom where the depth of water is 200 metres, or roughly 100 fathoms, we shall find that this line includes all these Isles, and that within it the sea is, with the exception of two or three isolated spots, as in the Channel between the coasts of Scotland and Ireland, and at the entrance to Loch Fyne and elsewhere, nowhere more than that depth[2]. If the sea bed round our land were raised 600 feet the water frontiers of the United Kingdom would disappear.

If we glance at a hydrographic chart of the North-Atlantic Ocean we shall find that the 2000 metre line forming the limit of the continental area is situated at a comparatively short distance from the coasts of Africa and Southern Europe, and is there at certain places no more than about 50 or 60 miles from the coast. Out from the coasts of Galway this line is very near the land and the ocean bed slopes very steeply down to abyssal

---

[1] By this term is to be understood not only the North Atlantic proper, but also the Norwegian Sea, the Arctic Sea, the North Sea and the Baltic.

[2] Off "Skate Island" in Lower Loch Fyne a sounding of 107 fathoms has been made (Mill, *British Association Handbook*, Glasgow Meeting, 1902). There is a depth of 149 fathoms in the deep "gutter" between Belfast Lough and Wigtonshire. Some of the Scottish inland Lochs (Loch Morar and Loch Ness), are also very deep. A sounding of 170 fathoms has been made in Loch Morar, which is only about 31 feet above sea level (Scott, *Ann. Rep. Fishery Board, Scotland*, for 1892, Pt. III. p. 221). The bottom of this fresh-water loch is therefore lower than any part of the North or Irish Seas.

depths. It is here that we find the extreme western boundary of the British submarine plateau. Out west about 200 nautical miles distant from the centre of Scotland is the Rockall Bank, an elevation of the bed of the Atlantic which rises up to about 200 metres from the surface. Between this and the British Islands there is a tongue of deep water which extends up to the latitude of Cape Wrath. The 2000-contour line of the Atlantic basin curves round to the south and west of the Rockall Bank and reaches to within about 15° S. from Iceland.

If we look again at the hydrographic chart we find that an extension of the continental shelf traverses the whole bed of the north Atlantic in a north-westerly direction and connects the British plateau with the coast of Greenland. Upon this are situated the Faeroe Islands and Iceland, and round these and off the south-west coast of Greenland are extensive "banks" or areas of shallow water. Between Iceland and Greenland is a wide extent of shallow sea; between Iceland and the Faeroes is a wide channel with an average depth of about 400 metres; and between the Faeroes and the British plateau is a comparatively narrow tongue of deep water—over 1000 metres in depth—which is the Faeroe Channel. In latitude 60° N. a narrow ridge connects the British and Faeroese plateaux. This elevation of the ocean bed is only a few miles broad, and over it the sea has an average depth of about 500 metres. It is the well-known Wyville-Thomson Ridge.

Thus a series of submarine elevations join the continent of Europe with the island of Greenland and separates the ocean into two basins, those of the North Atlantic, and the Norwegian Sea. On each side of the Wyville-Thomson Ridge the sea bed sinks down rapidly to depths of over 1000 metres. To the south of the ridge the sea slopes rather gradually down into the depression between the Rockall Plateau and the coasts of Ireland and Scotland. North of the ridge the sea bed slopes much more steeply down into the Faeroe Channel and then into the Norwegian basin. We shall see the significance of this disposition of the ocean bed later on.

It is on this shallow portion of the sea bed, practically in the sea within the 200-metre line, that deep-sea fishing is carried on.

In former times fishing was practically confined to the narrow strip of water within a few miles of the coast, and then gradually extended so as to include the North and Irish Seas. Then with the increasing demand for fish caused by the natural growth of the population, and by the increasing facility of transport, the fishing area was extended. As formerly, while yet no doubts were entertained as to our naval supremacy, the territorial limits of England extended over the seas to the coasts of the enemy's countries, so now at a time when we are still the predominant fishing nation of the world, we call the shallow water off the coasts of Iceland, Russia Norway, Denmark, Holland, France and the Peninsula the "British Fishing Grounds." British steam fishing vessels now exploit the coasts of Iceland in the west, the White Sea in the east, and the coast of Morocco in the south. Only a very few years ago trawling was confined to the sea which was not more than 100 metres in depth, but now the trawl is used in water which is more than double that depth, and in a few more years we may expect a further extension of the area within which fishing may profitably be carried on.

Now round the coasts of this extensive region the sea and the sea bottom present very variable physical and biological features. Fringing the land is a narrow strip of sea bottom which is bared twice a day by the ebb and flow of the tides. The area of the foreshore varies according to the nature of the coast and the depth of the contiguous sea. Off the coast of some parts of England it is very extensive and hundreds of square miles of sea bottom are twice in every 24 hours laid bare and again covered. Morecambe Bay off the coast of Lancashire, for instance, is such an evanescent sheet of water. The foreshore is the haunt of a host of creatures which form an abundant material for the fishermen. Gregarious mollusca and crustacea abound there, and at the time of high water innumerable shoals of fishes find their food in the shell-fish which are the characteristic fauna of the shore between tide marks. Outside the foreshore again is the zone of territorial waters—the sea within a line drawn three miles from low-water marks. This is the traditional fishing ground. Passing out from the foreshore the sea bottom in some parts slopes very gradually out into deep water. Off the coast of Lancashire the contour line of 20 metres

(10 fathoms) is reached only after we get 10 miles or more from the land. Even if we make a traverse of the North Sea between the Wash and the Texel on the coast of Holland, a distance of 150 nautical miles, we do not find water which is more than 10 fathoms in depth. In such parts of the sea the sandy foreshore is continued out as a flat sandy sea bottom underneath shallow water. In other places the coast-line descends more or less precipitously into the sea and we find a rocky shore covered with sea-weed. Passing out to sea from such a shore we soon encounter the "Laminarian zone," the portion of sea bottom covered with the giant fronds of the sea-weed *Laminaria*, and affording shelter for a varied assemblage of marine animals.

Sea bottom deposits. Outside the narrow region of coastal waters, the extent of which depends on the steepness of the continental slope, we come upon a sea bottom which is much less prolific in life than that which we have been considering, and the nature of which is very different. From the coast-line out to the edge of the continental shelf, at a depth of 200 metres, we have the region of the "terrigenous" sea bottom deposits. Near the land where the sea deepens rapidly, and over wide areas such as the North Sea in its southern portion, and the whole of the Irish Sea with the exception of the deep "gutter" between the coast of Ireland and the Isle of Man, the sea bottom consists of boulders, gravel and sand, all materials which result from the detrition of the land and the erosion of the coast-line. Sand and mud are carried in suspension by the water or are rolled along the sea floor by the action of tides and currents and are laid down evenly, forming flat areas of sea bottom with shallow inequalities or channels. Finer particles of mud are carried further and are deposited at a greater distance from the land. We find therefore that as we pass out beyond the edge of the continental shelf the bottom deposits become finer and finer until we reach in deep water the area of the terrigenous blue and green muds. But while such materials resulting from the waste of the land are, generally speaking, characteristic of the sea deposits within the 2000-metre line we yet find others which have a different origin. These are the "Benthic" deposits. While the terrigenous sea floor has in its chemical composition a preponderance of silica, this mineral constituting from one-half to two-thirds of its weight, these benthic deposits are characterised by their abundance in lime. Calcium carbonate constitutes from one-half to two-thirds of their weight. They consist largely of the broken down and comminuted remains of the skeletons of such animals and plants as form calcareous shells or skeletons. These are corallines or calcareous algae, corals, echinoderms such as starfishes, sea-urchins, &c., polyzoa, the spicules of calcareous sponges, and the shells of mollusca and crustacea. When these animals and plants die their remains form the sea floor, or at least a considerable part of it in some localities. In speaking of the terrigenous deposits which are characteristic of the sea bottom within the 2000-contour line we must include these benthic or "neritic" deposits which are not derived from the land, but from the bottom-living animals which inhabit this region to a greater extent than that underneath deeper water. (See Appendix.)

The term "pelagic" was first applied to the deposits of the ocean floor in the deeper parts of the ocean and at a great distance from land. Outside the 2000-metre line the nature of the sea floor is strikingly different from that within the continental area. While the remains of animals living on the sea bottom do indeed occur, these are not abundant and do not constitute a notable proportion of these deposits, and the only material composing them which has a terrestrial origin is such as has been derived from the material discharged during volcanic eruptions. Pumice from such a source finds its way into the sea directly or after carriage by rivers, and after floating on the surface of the sea becomes water-logged and sinks to the bottom there to undergo decomposition. Volcanic ash too is carried in the atmosphere and finally may fall on the sea and sink. These substances occur in the muds or oozes which form the floor of the deep oceans, but mud or sand derived from the detrition of the land is not found there. Instead we find a material which is almost entirely derived from the remains of the animals and plants which live in the body of sea water extending from the surface down to the bottom.

At a limited distance from the land and still within the limits bounding the area of the terrigenous deposits may be found the

Pteropod Ooze, a substance which is characterised by the preponderance of the shells of Pteropods, pelagic Gasteropoda which live at the surface of the sea in great shoals. From about 1000 to 3000 metres is the depth of sea in which Pteropod ooze is found,

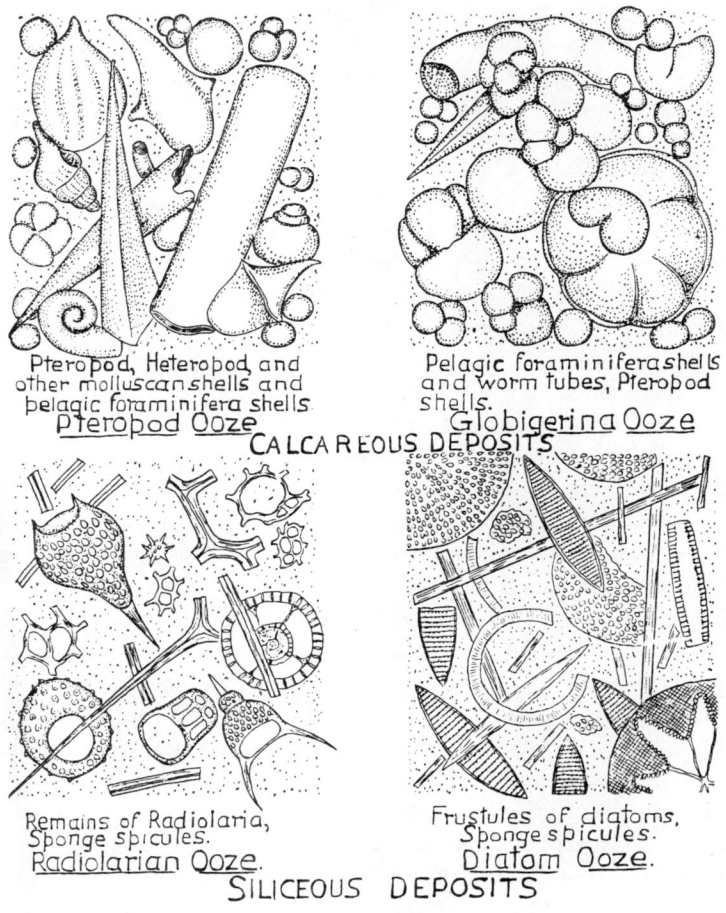

Fig. 14. Sea bottom deposits.

outside these limits is the area of distribution of the well-known **Globigerina Ooze** which is characteristic of a greater extent of ocean bed than any other deep-sea deposit. It is formed of the calcareous skeletons of the numerous species of foraminifera which inhabit the upper layers of the sea. Its average range in depth is from about 3000 to 5000 metres and at those depths Atlantic and Norwegian Oceans are alike carpeted by this deposit.

But beyond a depth of about 5000 metres the remains of organisms possessing a calcareous skeleton form but an insignificant proportion of the oozes on the ocean floor. It takes so long a time for these fragile remains to fall through water of this depth that before reaching the bottom they are almost entirely dissolved. Siliceous deposits take their place. Of such nature is the **Radiolarian Ooze** which is found at greater depths than 5000 metres. The Radiolaria are Protozoa which secrete a siliceous skeleton, and which inhabit nearly all layers of the sea. When they die their bodies sink to the bottom and while the soft parts decompose slowly and disappear the siliceous remains accumulate to form a deep-sea deposit.

Beyond the limit of depth of the radiolarian ooze we find the **Red Clay** which lies in the great abysses of the ocean. While some part of this deposit is formed by the remains of living creatures it is nevertheless a substance which is mostly formed from inorganic material. We have seen that pumice and volcanic ash may be carried over wide tracts of ocean surface and that this material finally becomes water logged and sinks to the sea bottom. This is the origin of the red clay. At the profound depths in which it is found almost all organically formed material has disappeared from the deposits, being dissolved by the sea water or perhaps broken down by the action of marine bacteria—a mode of action which conceivably may account for the disappearance of much of the silica and lime which must reach the sea floor from the upper layers as the skeletons of marine organisms.

In addition to these substances composing the oceanic bottom deposits there are of course many others. None of them are pure, but each shades off into the others under various conditions and at various depths. Other formations also occur, such as the well-known **Diatom Ooze** of the Antarctic. At an average depth of about 4000 metres there is a wide band of soft white ooze covering the sea bottom between the parallel of 40° S. and extending to the Antarctic circle and completely surrounding the southern hemisphere. This band has an area of over ten millions of square nautical miles. The material composing it has resulted almost entirely from the siliceous shells of diatoms which live in enormous numbers at the surface of the antarctic seas, and which dying, sink to the bottom, where their skeletons accumulate. Then too in all these oceanic oozes other substances are found. All over the deep basins of the oceans there are found the teeth of sharks and ear-bones of whales, these being the only parts of the skeletons of these creatures which withstand the solvent action of the sea water at the great depths in which they are found. In the Norwegian Sea there occur the dense calcareous otoliths or ear-bones of various teleostean fishes. All over the sea bottom are found the peculiar manganese nodules which were first described during the famous voyage of the *Challenger*. Finally it has been recognised that in the deepest parts of the sea there are particles in the oozes which have an extra-terrestrial origin: particles which result from the combustion of meteorites entering the atmosphere from without. These should of course occur in the deposits of all depths, but in most their presence is masked by the preponderance of other substances, and it is only in the deposits of the very deep basins, which are formed with extreme slowness, that they can easily be recognised.

In the North-Atlantic Ocean the **temperature of the water** varies from place to place very notably, these variations being determined by a very definite circulation of the water on the surface and at some depth beneath this. A glance at a chart shewing the surface isotherms, such as will be found in the *Challenger Reports*, will illustrate this variation more clearly than any amount of description. It will be seen that the mean annual temperature of the surface varies from about 4° C. at the Arctic circle to about 29° C. in the region of the equatorial stream (Lats. 0° to 30°). Further it will be seen that the isotherms do not run in a direction parallel with the lines of latitude, but are bent north and south in a way which is at first very puzzling but is very easily explained by a consideration of the facts of the water circulation. Roughly speaking there are three wide belts or regions, (1) the region of the equatorial stream where the mean annual temperature of the surface is about 29° C., (2) a wide irregular band between the equatorial stream region and latitude 47° where the mean annual temperature varies from 29° C. to 15·5° C., and (3) a region extending from the northern limit of (2) to within the Arctic circle where the mean annual temperature varies from 15·5° C. to 4·5° C., this latter area including the British Islands. It will also be apparent that the isotherms do not run parallel with the lines of latitude, even roughly in such a way as we see is the case in south temperate seas, but north of latitude 40° are bent to the north-east, and south of this parallel are less strongly bent to the south-east. This disposition is due entirely to the influence of the Gulf Stream and the north-easterly drift of Atlantic water, and to the much less abundant southerly flowing polar stream which is evident on the shores of North East America.

In the North Atlantic, in the axis of the Gulf Stream drift W. from Ireland, the annual range of temperature is 5° C. This is naturally much less than is observed in the shallow seas such as the North, or Irish Sea, where the cooling and heating of the land, and the entrance of water from rivers, have a great effect on the temperature of the sea. In the North Sea we have extreme yearly variations of 2° C. to 17° C., and in the Irish Sea of 3° C. to 18° C. Where there are shallow shores, and strong tides, the sea off the land may be strongly heated or cooled; for as the foreshore is laid bare by the tide it is heated or cooled, and water flowing over it is distributed by the tidal currents.

Quite another series of variations are to be observed if we study the vertical distribution of the temperature of the sea. This may be done by making "hydrographical soundings" in the manner described in the last chapter. We have seen that an elevation of the ocean bed extends across from Scotland to the Island of Greenland, and this ridge forms a "barrier" which is of the greatest significance in the vertical distribution of the temperature of the sea. On one side of this barrier the sea slopes down into the depths of the Atlantic Ocean, and on the other it deepens to form the basin of the Norwegian Sea. Within comparatively short distances of the Wyville-Thomson Ridge we find that the Atlantic has a depth of 3000 metres, while in the Norwegian Sea there is an extreme depth of about 3700 metres. Water currents tend to flow along deep depressions and we find that the Wyville-

CH. II] THE OCEANOGRAPHY OF THE NORTH-WESTERN OCEAN 39

Thomson Ridge acts as a barrier to the flow of warm water from the Atlantic into the Norwegian Sea, and conversely to the flow of cold water from the Norwegian Sea into the Atlantic. Fig. 15 represents a series of hydrographic soundings on the surface of the

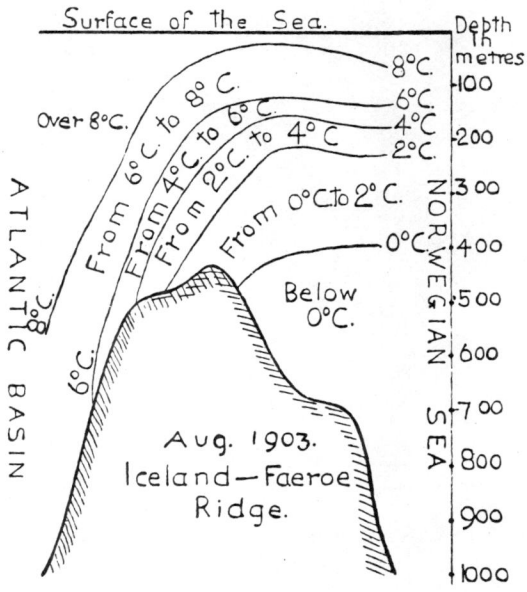

Fig. 15. Hydrographic section over the Wyville-Thomson Ridge, in the direction of the axis of the Faeroe Channel. (From *Rapp. et Proc.-Verb.; Cons. Perm. Int. Explor. Mer*, Vol. III. p. 6, Appendix G, 1905). Shews the distribution of temperature.

ridge and on either side, and shews the effect of such a barrier on the flow of an ocean current. At the surface of the sea the water temperature is 8° C. to 11° C., and at the bottom on the top of the ridge it is from 2° C. to 4° C. But it will be seen that whereas on the Atlantic side the temperature at a depth of 450 m. is 8° C., on the Norwegian side at a corresponding depth the temperature is 0° C. At this place we have a double flow of water: at times cold water from the Norwegian Sea tends to pass over the ridge from the north, while Atlantic water at a much higher temperature is continually passing over it from the south. But only those layers which are above the level of 400 m. from the surface pass over the ridge.

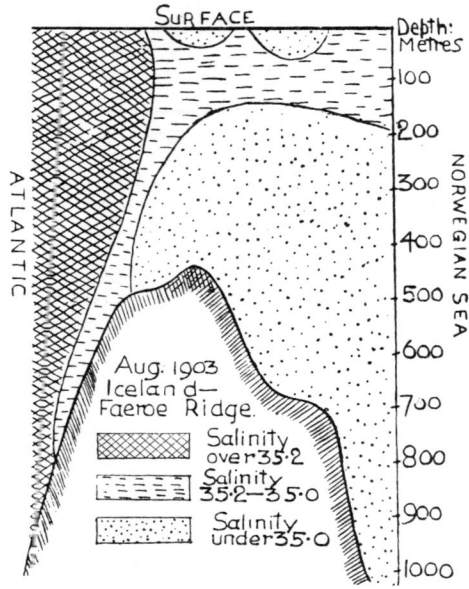

Fig. 16. Similar section. Shews the distribution of water of various degrees of salinity.

On the Atlantic side the temperature of the sea is highest at the surface and then falls rapidly for the first two hundred metres until we reach a kind of boundary between the variable warm water at the surface, which is considerably influenced by the seasons, and that lying on the sea bottom. From this level the sea cools less rapidly down to a depth of about 1000 metres and then it cools very slowly down to the bottom where there is a temperature which does not vary and is about that of the freezing point of fresh water. On the Norwegian side the conditions are very different. There is a surface stratum which is about 600 metres from the surface and in which variations may occur, the temperature ranging from 1·5° C. to about 8° C., but below this the water has a uniform temperature which is about 0° C. while in the deepest parts it is as low as − 1·3° C., lower than any other part of the ocean bed of the world with the exception of a place in Bransfield Sound in the Antarctic where a temperature of $-1\cdot6°$ C. has been observed.

Colour and transparency. Over this area another physical difference is that of the varying distribution of light both at the surface and in the deeper layers of the sea. If we attach a white disc, such as an enamelled iron plate, to the end of a sounding line and lower this into the sea we find it disappears at a variable depth, which depends for the most part on the amount of mud or fine inorganic particles in suspension in the water. Where these are not present differences in the transparency of the sea depend on the colour of the water and on the amount of the planktonic organisms in it. In the turbid waters off the coast of Lancashire such a disc may disappear at a distance of a few feet if we make the experiment during the time of strong spring tides, or at a time when strong winds are prevalent. Even in the most favourable circumstances, that is during fine weather, in brilliant sunlight and at some considerable distance from the land, it is seldom that a white object can be seen at a greater depth than 20 metres (10 fathoms). In the Lake of Geneva a lighted electric lamp can be seen at a depth of 38 metres, in the Pacific Ocean a white disc may be seen at 40 metres. The most transparent waters are those of the Mediterranean, where the maximum translucency of the water is estimated at 45 metres. With these differences in transparency there are also differences in colour. The Baltic is dirty yellow, the North Sea is greenish and the open seas are green to blue.

Fol and Sarasin investigated the penetration by light of the water of the Lake of Geneva by means of photographic plates. These were enclosed in shutters and lowered into the water to variable depths, and the shutter was opened by means of a "messenger" or some similar contrivance. It was thus found that at a distance from the surface of 170 metres the plates were affected, but it was estimated that the intensity of the light could not be greater than that of a dark moonless night. Similar experiments carried out in the Mediterranean gave a depth of about 400 metres as the limiting one: below this light, as the human eye perceives it, is entirely absorbed by the overlying layers of sea. We may conclude then that at an average depth of about 200 metres profound darkness reigns. But this does not exclude the possibility of some degree of perception, by abyssal animals, of radiations from the atmosphere, although this may be of such a nature that our eyes would not be affected by it.

At the bottom of a deep sea uniform conditions obtain. The bottom is a flat plain with few inequalities, for those indicated by the sounding machines are slight compared with those we know on the land, and though precipitous declivities must occur these are very exceptional. The sea bottom is composed of soft semi-fluid oozes into which objects must easily sink. A uniform temperature which is that of the freezing point of fresh water, or only a degree or two above this, obtains. Profound or absolute darkness, broken only by the light of some phosphorescent creature is there. Daily or seasonal changes never occur and almost absolute uniformity of conditions reigns. Add to these the enormous pressure of the overlying water, which is about one atmosphere for every ten metres of depth, and we have conditions in which it is almost incredible that life as we know it can exist.

The water circulation. To complete this sketch of the oceanography of the North-Western Ocean we have now to consider the circulation of the sea water. We have seen that temperature observations give indications of considerable movements of the waters of both the Atlantic and Norwegian oceans. It is indeed by indirect methods that these have been traced. Direct observations of the movements of the water both at the surface and at the bottom have been made, and many ingenious forms of apparatus have been devised for this purpose. But the main facts of the circulation of the North-Atlantic Ocean have been made out by practically simultaneous observations of the temperature and salinity of the sea currents over wide areas. A knowledge of these two important characters of sea water enables us to determine with some degree of certainty from what area the water has been derived. The salinity is the total amount of dissolved salts per litre of water: it is a number which varies from 37 in the subtropical Atlantic to eight in the Baltic: that is to say, water in the region of the Atlantic Equatorial Stream contains about 37 grams of dissolved salts in 1000 grams, while in the comparatively fresh water of the eastern Baltic this weight of water contains only eight grams of total salts. Generally speaking the salinity of the Atlantic south of the Wyville-Thomson Ridge is over 35, and a similar value holds for the surface of a considerable part of the Norwegian Sea. In the North Sea the mean salinity varies from 35·2 to 34·6 and is highest at the entrance to the Straits of Dover in the south and between the Shetlands and Norway in the north[1]. In the North Sea in a line drawn from Grimsby to Texel the salinity is lowest, that is 34·5. In the Irish Sea the values for the salinity vary much as in the North Sea.

Three components enter into the composition of the water of the North-Atlantic Ocean: (1) "Gulf Stream" water, (2) cold bottom water from the Norwegian Sea, and (3) fresh water from great rivers. Extensive water bodies retain their heat for a long time, and diffusion in liquids being a slow process, the mixture of different layers takes place very slowly. We find then that over wide areas the three components mentioned above may be traced with more or less distinctness. The density of sea water depends on the temperature no less than on the salinity. Thus a water layer at a high temperature and having a high salinity may still be specifically lighter than a water layer at a low temperature but with a low salinity. This is the case with the Atlantic stream which enters the Faeroe Channel. It is really richer in dissolved salts than the water normally belonging to the Norwegian Sea, but since it has a higher temperature than the latter it is lighter on that account, and so it flows on the surface of the cold fresher water until when it has cooled down it sinks beneath the surface and still flows on as an undercurrent.

If a great number of observations of the temperature and salinity of the sea over a very considerable area be made at about the same time, and if these simultaneous observations be repeated at frequent intervals, say every three months, then a picture of the water circulation in the area may be made with some degree of

[1] A chart shewing the mean salinities is given by **Martin Knudsen** and **Miss K. Smith** in *Rapports et Procès-Verbaux; Cons. Perm. Int. Explor. Mer*, Vol. VI. 1906.

probability. For liquids of different density and temperature diffuse into each other so slowly that an extensive area of sea may consist of roughly defined streams or strata which only mix slowly at their boundaries and preserve their individuality over a considerable extent of sea surface. A comparison of such results as they are indicated in the charts published quarterly in the *Bulletin des Resultats* of the International Council for the Exploration of the Sea will shew that considerable variations in the extent of the sea surface covered by these different layers occur from time to time. These are indicative of changes in the situation of the water components of the Northern Ocean, that is to say they are evidence of a water circulation in this area. Similar evidence is furnished by a study of the microscopic animals and plants present in the water, since these are characteristic of the place of origin of the stream.

We have seen that Gulf Stream water, Arctic water from the Polar Sea and fresh water from the great rivers of the land all enter into the composition of the seas of Northern Europe. The first of these components is by far the most important. It is well known that the shores of the British Islands are bathed by water of subtropical origin, usually called "Gulf Stream" water, and the statement that the climate of these islands is much milder than it would otherwise be is a well-worn one. It is quite true that water of subtropical origin, relatively warm and dense, does reach these latitudes from the Atlantic Ocean, but for a long time it has been known that this water is not an offshoot of the Gulf Stream.

The first complete investigation of the waters of the North Atlantic with reference to the distribution of the waters of the Gulf Stream was carried out in 1899[1] when the variations in the temperature, salinity and plankton of the waters of the Gulf Stream region were investigated from the point of view of the distribution of the waters of this famous current. It is shewn that the limits of the Gulf Stream were variable but that at no time in the year did it reach the British Islands. If we consider the isohalines (lines of equal salinity of 36 to 37) we find that these vary with the season. In March they reach no further north than the Azores, and then the sea between these islands and the coast of

---

[1] Cleve, Ekman et Pettersson, *Les variations annuelles de l'eau de surface de l'océan Atlantique.*

Africa is filled with water of less density and of lower temperature. From March to November the Gulf Stream area expands greatly in a north-easterly direction and in the latter month has approached the coasts of North Africa and Southern Europe, and at the same time there has occurred an expansion of the area over which tropical plankton may be found. Then from November to the following March the stream regains its former volume. Issuing from the Gulf of Mexico the Stream extends out into the North Atlantic in a north-easterly direction but curves round again to the south, forming a closed eddy of gigantic proportions. In the centre of this eddy is the Sargasso Sea, a streamless region filled with water of high temperature, and on the surface of this float sea-weeds among which are to be found fishes and other marine animals.

In some way or other a drift or current of water takes origin just north of the Gulf Stream eddy and this is the stream which reaches the shores of Great Britain and Scandinavia. Towards our latitudes the salinity and temperature decrease but the former is not less than 35·5, and off the western coasts of Great Britain the mean annual temperature is from 10 °C. to 13° C. It is often suggested that the cause of this drift is the prevalence of south-westerly cyclonic storms which reach our islands, and while this is no doubt a contributing cause of the movement of the water it is also probable that complex thermodynamical causes are at work. But however this may be it is the case that a conspicuous drift of water from the subtropical Atlantic bathes the shores not only of Great Britain and Scandinavia but also those of Iceland, and fills up the deeper parts of the Baltic and Barentz Seas. This is the European Stream.

Now notice how the flow of this current is restricted by the configuration of the ocean bed. The stream flows along the deeper channels and depressions and fills up all the sea to the south-west of the British Islands. It enters into the Bay of Biscay and from there flows towards the English Channel and through this into the North Sea. The main bulk of the stream then flows to the north-east and impinges on the submarine barrier which is formed by the Wyville-Thomson Ridge and by the Faeroese-Icelandic banks. South-west of the British Islands the Atlantic from the surface to the bottom is filled with water which is relatively heavy, and a very deep layer from the surface downwards is comparatively high in temperature. But encountering the Scotland-Iceland barrier the lower part of the stream is arrested and only the superficial layers flow on over the colder waters of the Norwegian Sea. Just such another barrier exists in the barrier which is present at the Straits of Gibraltar. This rises to comparatively near the surface and arrests the flow of Atlantic water of deep origin. Only superficial Atlantic water enters the Mediterranean, flowing over an undercurrent of warm water which passes out into the Atlantic. Down to great depths the temperature in the former is 12° C. to 13° C.

A glance at the chart on p. 47 will shew the further course of the European Stream. This is the first chart which was constructed from simultaneous salinity observations in the Atlantic the North and Norwegian seas, and it appears that it is not quite typical of the conditions usually occurring. It shews that Atlantic water enters the Norwegian Sea over all the banks between Shetland and Iceland. It is the case that the stream flows as a very general rule only through the Faeroe Shetland Channel In 1896 we have what is apparently a maximum extension of the Atlantic flooding of the Northern Ocean. It will be seen that warm and high-salinity water covers a large portion of the Norwegian Sea. It has encroached on the ice barriers round Greenland and Spitzbergen; has flooded the western entrance to the Denmark Strait; and finally loses itself in the Arctic ocean as far north as Lat. 75°, and as far east as Long. 50°. It has entered the North Sea as a surface current, and the Baltic as an undercurrent.

Now remember that the area covered with the waters of the Gulf Stream undergoes periodic expansion and contraction varying between 45° and 50° N. Lat. and that water of comparatively high salinity and temperature must be banked up in the Atlantic Ocean at these latitudes, displacing water which was previously there. Remember too that this gigantic pulsation of the Gulf Stream occurs once a year; that the maximum expansion of the eddy occurs in November and minimum contraction in March. It must follow from these conditions that the volume of the European Stream, which depends on the banking up in the Atlantic of water from more southerly latitudes, must vary with the season. This is actually the case. There is indeed a uniform flow of Atlantic water over the Wyville-Thomson Ridge—uniform in the sense that it is continuous—but if we study the appearance of the

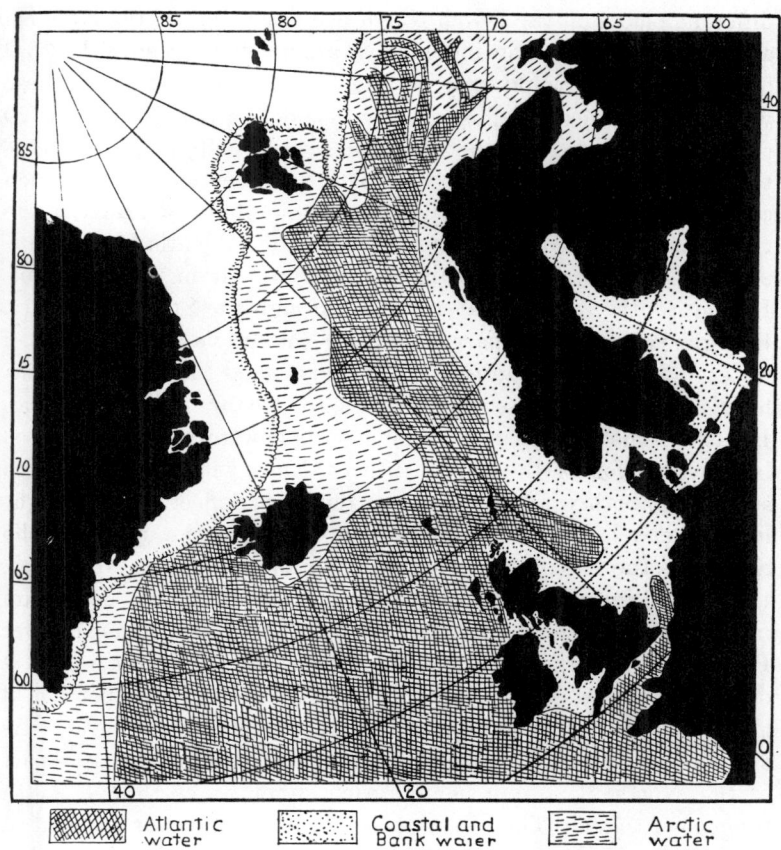

Fig. 17. The North-Western Ocean shewing the European stream and its offshoots. From Pettersson, *Rapp. Proc.-Verb.* Vol. III. 1905.

stream in its remote ramifications, as in the North Sea, the Skagerak, or in the Barentz Sea we see that yearly pulsations, analogous with those which occur in the Gulf Stream eddy, do really take place. The most curious proof of the correspondence of the annual flooding of the Northern Ocean with Atlantic water is to be found in the annual rise and fall of the waters of the North Sea and Baltic. It is possible to dissociate the rise and fall of the sea due to this cause from the fluctuations in level due to tides, winds, &c. When this was done the water gauges at nine different places on the Dutch North Sea coast, and on the Swedish and German Baltic coasts indicated an annual fluctuation of the level of the sea which could only be due to a yearly flooding of these waters. The rise and fall is in every case only a matter of about 20 to 26 centimetres, but the almost exactly synchronous nature of the variations shewed that they were due to a common cause, which can be no other than the variable volume of water which enters these seas in the course of the year from the Atlantic Ocean.

The North Sea receives high-salinity water from two directions. In the South Atlantic water enters through the Straits of Dover. But this is only a very narrow passage and only a small volume of water enters here. The principal flow is round the north of Scotland. After passing through the Faeroe Channel the Norwegian Branch of the European Stream, as it is now termed, is deflected by the rotation of the earth, in accordance with the law which states that both wind and ocean currents are deflected to the right in northerly latitudes. Passing then round to the north of the Shetlands the stream turns to the south and enters the North Sea. It grows in intensity during the winter and reaches its maximum in the spring, afterwards decreasing in strength. The Dogger Bank in the northern half of the North Sea interrupts the flow of the northerly Atlantic stream so that the southern portion of the area receives its dense water from the Straits of Dover. As the spring advances these two currents gradually cover a large part of the North Sea, approaching each other as north and south tongues of water. They do not however meet, and over the central area of the North Sea the Atlantic water is mixed with fresher water arising from the Rhine, Scheldt, Weser and Elbe, or perhaps coming from the deep Norwegian channel into which it flows from the Baltic. But undercurrents may pass along the deeper depressions and guts on the English side and join the north and south Atlantic tongues. It has been estimated that such undercurrents may attain a velocity of about five kilometres per day

It will be seen from the chart of the North-Atlantic ocean that a deep depression lies just off the coasts of Norway and Sweden. This leads into the Skagerak. The Atlantic water coming round the north of Scotland flows into this channel and becomes banked up there by the shoaling of the bottom. As summer advances this water becomes covered up by lighter and warmer water, so that at the bottom the temperature is from 4° to 7° C., while on the surface the temperature is about 16° C. In the February following, while this bottom water still retains its heat, the surface has cooled down to 2° to 3° C. An undercurrent from this heaped-up bottom stratum then enters the Baltic, where it produces momentous effects on the course of the fisheries.

Finally the Norwegian Branch of the European stream reaches as far as the remote waters of the Barentz Sea, and enters that area as the so-called North Cape current. "The geographical position of the North Cape current and its ramifications is constant, quite as constant as is that of our rivers[1]."

Annually the Barentz Sea is invaded by a heat wave which is the result of the increase in the volume of the Atlantic water which enters it. The summer of the sea is November; its winter is June. In the latter month the temperature of the bottom layer is 1° C., in November it is 5° C. or more. This increase in temperature is due to the increased flow of the Atlantic stream, which attains its maximum in the latter month. In the Barentz Sea the stream has cooled down so that it now flows on as an undercurrent. With the Atlantic flow fishes like the cod and haddock invade the Barentz Sea. In November the stream of genial water begins to decrease and cold Arctic water takes its place. With this change the fisheries cease.

We see then that there are three principal factors in the circulation of the water of the North-Atlantic Ocean. There is first the European Stream which flows up from the Atlantic Ocean south-west of the British Islands. Then we have a much smaller stream of cold water which flows down from the Arctic as the East-Icelandic Polar Stream and which may at times enter the Faeroe Channel and obstruct the flow of the European Stream.

---

[1] Breitfuss, "Oceanographische Studien über das Barentz Meer," *Petermann's Mitth.* 1904.

Lastly we have a considerable volume of fresh water entering the North Sea from the great rivers of the Continent, and an outward flowing current from the Baltic Sea which consists of water which is always fresher than the water of the sea into which it flows.

Now the strength of these various currents is not always the same. We have seen that the extent of ocean covered by the Gulf Stream eddy varies from time to time throughout the year, being greatest in November and least in March. This is a periodic variation which depends on the shifting of the equatorial currents as the earth travels round the sun. Because of this variation of the area covered by the Gulf Stream eddy there is in November a greater amount of warm water banked up in the temperate Atlantic than in March, and as a consequence the volume of the European Stream in northerly latitudes is greater in the early spring than in the late autumn. This is the periodic variation of the Atlantic flooding of the northern seas. So also we have a periodic variation of the Polar current, less marked, it is true, than that we have been considering, but still considerable enough to complicate the hydrography of the Northern Ocean. Lastly we have variations in the amount of water entering the North Sea from rivers, and variations in the strength of the Baltic current, these two latter obviously depending on the nature of the seasons. Year by year these various currents wax and wane, and we expect that each of them attains its maximum strength at a definite time of the year.

But there are variations in the flow of each of these currents which are not easily to be explained. In the case of a planet, the motions of which may be calculated according to our knowledge of various laws, there are found perturbations which are caused by influences which have not been taken into account. Just so in the case of these various currents, the strength and course of which may be expected to recur periodically from year to year, there are perturbations which can only be explained by assuming the existence of some influences of the nature of which we know next to nothing.

When we remember that the study of the oceanic circulation of the North-Atlantic Ocean by reliable methods dates back only from the beginning of the International Fishery Investigations in

1902 we will see how hopeless it is at the present time to attempt to describe the nature and extent of these perturbations, or unperiodic variations, much less to explain their cause by reference to general laws. For this purpose the accumulated observations of many years will be necessary. What we do know at the present time only gives us elusive glimpses of the working of great causes, probably cosmic in their nature. The maximum flow of Atlantic water into the Norwegian Sea ought to take place at the same time every year if we had to deal only with such a variation in the extent of the Gulf Stream eddy as would be caused only by the revolution of the earth and the inclination of the axis of the latter to the plane of the orbit. But we know that the time of maximum flow of the European Stream through the Faeroe Channel may vary by a month or more. Thus the study of the temperature variations of the sea at the Faeroe Islands shows that the Atlantic flow culminated in the year 1902 in September instead of August, which is the month in which this should normally take place. There is in fact no absolute fixity in the time of appearance of the height of the European Stream. Temperature observations taken on the coast of Norway show that there is a two-yearly period in the variation of the temperature of the sea. In the even years the temperature is above the normal, while in the odd years it is below the normal. These variations can only be due to the fact that the flow of comparatively warm water into this area from the Atlantic is less in the odd years than in the even; that is, the Atlantic Stream has a two-yearly periodicity. What is the cause of this? Another striking fact is mentioned by Pettersson in the paper quoted. It is now beyond question that the migration of the herring into the various fishing grounds of the North Sea depends on the presence of a layer of water of a certain density. That is, herrings are only present in the sea in sufficient numbers to form a fishery if the hydrographical condition of the water is favourable to them. But the temperature and salinity of the sea depend mainly on the strength of flow of the European Stream, in our latitudes of course. Now since the year 895 the winter herring fishery in the Skagerak has been recorded, and it is found that it has returned at intervals of, on the whole, 111 years. This is one of those isolated observations which just allow us to see, in a most imperfect way, the working of some complex natural law. The door is opened, so to speak, and then closed in our faces. That the cause of this "secular periodicity" is a cosmic one we can scarcely doubt; but what is it?

Climate and hydrography. We will consider later on in what manner the hydrographical conditions of the sea influence the abundance and migrations of fishes and other marine organisms, but in the meantime some notice may be taken of the connection between hydrographical and climatic phenomena. Every year the European Stream brings into northern seas an immense mass of water, the temperature of which is greater than that of the water which is normally present in those areas. An incredibly great amount of heat is thus annually yielded up to the atmosphere in these latitudes. The moist and temperate climate of the British Islands is the result of the oceanic circulation we have been considering. If the North Atlantic were a streamless ocean the isothermal lines would be roughly parallel to those of latitude; the sea round the British Islands would have a mean temperature equal to that of the sea off the coast of Labrador; and the climate of the land would be changed in a similar manner. But the isothermal lines are not parallel to those of latitude, but are bent up strongly to the north-east; the mean temperature of the sea off the coast of Labrador varies from 2° C. to 5° C., while that of the sea off the coasts of Great Britain is 10° C. to 13 °C.; and the temperature and climate of the two countries vary accordingly. These conditions are the result of the distribution of the European Stream which not only conveys warm water to the seas of the British Islands but restricts the cold Labrador current to the shores of North America.

A well-known meteorologic phenomenon is the retardation of the seasons in the countries surrounding the Norwegian Sea, the North Sea and the Baltic. In the eastern part of Europe, and on the coasts of Scotland, of Ireland, and on the south and west of England, the coldest month of the year is January. In the Shetlands, the Faeroes, the Scandinavian Peninsula, &c., the season is retarded and the coldest month is February. This variation in the incidence of the seasons is ascribed to the fact that a great amount of heat is stored up in the water accumulated in the basins of the seas mentioned.

If then, as is undoubtedly the case, the climate of the maritime countries of Northern Europe is conditioned by the carriage of heat in the water coming from the Atlantic Ocean we should expect to find that climatic perturbations would result from variations in the strength of the European Stream. The delay in the time of culmination of the current must mean the delay in the winter cooling of the sea and consequently to some extent in the cooling of the land. The "odd and even years" are familiar to meteorologists. The odd years have been, as a rule, colder in the winter than the even ones. We have seen that the temperature of the sea is lower in the winter during the odd years than during the even ones, and variations in the temperature of the sea are analogous with those on the land. Other climatic phenomena, into a consideration of which we have not space to enter, are connected with the variations in the strength of the European Stream. The unusual accumulation of abnormally warm water in the sea round Iceland and off the coast of Norway and Spitzbergen delays the southern encroachment of ice floes, and serves as the breeding ground for cyclonic disturbances by producing abnormal conditions of atmospheric pressure. Obviously the temperature of the wind varies according as it blows over a warm or cold sea surface. This is well illustrated by Knudsen in a discussion of the influence of the East Icelandic polar stream on the climatic changes in the Faeroe and Shetland Isles and in the north of Scotland[1].

The main facts of the circulation of water in the North Atlantic basin are those elucidated by the hydrographic work of Pettersson and Cleve, and this statement is based on the published work of these investigators. Obviously the data are susceptible of more than one explanation, and since the hydrographic investigation of the north-western ocean is still proceeding it is not unlikely that Pettersson's theory, which I have followed in this place, may receive some modification in the future.

[1] *Rapports et Procès-Verbaux*, Vol. III. 1905.

CHAPTER III.

LIFE IN THE SEA.

IN 1839 Edward Forbes made the first attempt at a rational classification of marine organisms with respect to their physical surroundings by suggesting that four clearly defined zones could be distinguished round the shores of the British Islands, in each of which there was a characteristic fauna and flora. Between high and low water tide marks was the Littoral Zone, a region daily exposed to the atmosphere by the withdrawal of the tide. Then followed the Laminarian Zone, so called from the predominance of the large leathery-looking sea weed, *Laminaria* or tangle: this extended down to about 10 fathoms. Below this again was the Coralline Zone, a region in which the ordinary sea-weeds began to disappear and in which the nullipores, that is the Algae with calcareous skeletons, were the characteristic organisms. Beyond this again was a region extending down to an unknown depth in which vegetable life was entirely absent and in which animal life gradually began to disappear. Forbes thought that as we descend the ocean depths to the abysses organisms become more and more modified until in the great depths of the sea life became extinct, or exhibited only "but a few sparks to mark its lingering presence."

This speculation was made prior to the great oceanographical discoveries of the middle of the nineteenth century. Darwin had just returned from his *Beagle* cruise, and Sir James Ross was about to sail to the Antarctic with the *Erebus* and *Terror*. But even then Sir John Ross had returned from the great Arctic voyage of 1818, and by the successful soundings which he had made in water of over 1000 fathoms, and by the use of the "Deep-sea Clam," had shewn that life existed even at that great depth. In 1839 the dredge—" an instrument as valuable to the naturalist as a thermometer to the natural philosopher"—was being widely used, but the apparatus for employing it in deep water had not yet been elaborated. Forbes based his speculations on the results of his own dredging, which was made in water which we should now call "shallow," and since, like all men of an original turn of thinking, he reasoned from the observations made by himself, he so fell into the error of drawing large conclusions from insufficient data. Although he stated his beliefs hypothetically they were accepted widely and stated by others as dogmatically as *à priori* speculations of the time could be; and for a long time afterwards the deep sea was regarded as lifeless because of the apparent impossibility of animated things living under the influence of immense pressure, and in the total absence of light and air. Also the zones of distribution suggested by Forbes were accepted by reason of their practical convenience, and are indeed still referred to in faunistic discussions. But the great voyages of the middle of the century and after soon extended our knowledge of the distribution of marine animals with respect to the depth of the sea, and it was by-and-by conclusively proved that there was no region of the sea bottom which was entirely devoid of life.

In 1839 the dredge was the instrument of investigation of the marine naturalist, and knowledge of marine life was practically restricted to the animals which lived on the sea bottom and could be captured by means of this apparatus. The fishes and other larger animals living in the upper layers of the sea were, of course, known, but the incredible abundance of marine life which the tow-net and microscope were to reveal was then still unsuspected. Not until 1845 did Johannes Müller, the great anatomist and physiologist, begin to study the pelagic life of the sea by means of the examination of samples of sea water. This method of work was soon superseded by the use of the tow-net, and the material collected by this instrument at once afforded an inexhaustible field for systematic zoological investigation, and for the study of the development and life histories of the previously known bottom living animals. Huxley, Haeckel and many other zoologists began systematically to use the tow-net, and by the employment of this apparatus on board the *Challenger* in 1873–6 an enormous advance was made in our knowledge of marine life.

When to the creatures captured by the dredge and the apparatus of the fishermen were added those obtained by the tow-net it became possible to devise a new grouping of the organisms found in the sea. In 1887 Victor Hensen published the results of a voyage of exploration in the North Sea and Atlantic Ocean, and in this now classical memoir (*Ueber die Bestimmung des Plankton's oder des im Meere triebenden Materials an Pflanzen und Thieren*) he used for the first time the now familar word *Plankton*, replacing by it the older " Auftrieb " of Johannes Müller. In 1890 Ernst Haeckel published the famous *Plankton-Studien*, a memoir largely controversial in treatment, but containing the *résumé* of a wide experience in the investigation of the plankton of the sea. In this paper Haeckel invented a number of new terms to be used in such discussions terms which have not all met with general acceptance, but some of which have been adopted. In the *Plankton-Studien* Haeckel proposed to distinguish the animals which are caught in the tow-nets from those which are obtained generally by other means. If we take a broad survey of organisms found in the sea and consider their habit of life we will find that all can be grouped in three great categories. There are first of all those which by reason of their minute size and feeble powers of locomotion are carried about passively in the sea by tides and currents. These are they which are caught in the tow-nets, which Müller called the Auftrieb, and Hensen the Plankton. Then there are those animals and plants which can be taken by means of the dredge or gathered on the foreshore. Sea-weeds, molluscs starfishes and their allies, zoophytes and most crustacea belong to this group. They are organisms which, like the sea-weeds and the zoophytes, live attached to the sea bottom, or like the molluscs and echinoderms live there more or less permanently and make few or unimportant migrations. These form the " Benthos " and Edward Forbes' classification included only such creatures. Then opposed to both these categories are the numerous class of animals which roam about over comparatively wide areas of sea. Such are the fishes, the marine mammalia and some molluscs and crustacea. For the reception of these animals Haeckel, to whom is due also the term benthos, coined the word "Nekton," and the term has come into general use. There is, of course, no absolute distinction between these three classes of organisms. That is, there are creatures which, like the molluscs and crustacea, belong in their adult phases to the benthos, but which in the larval state live among the plankton; while on the other hand there are some animals, such as the fishes, which are inhabitants of the nekton but which are also in the young state to be reckoned among the plankton. Then one at times finds it difficult to say whether organisms, like the medusae, which are carried about in great swarms by tides and currents, but which nevertheless are capable of some degree of locomotion, are to be included in the plankton or in the nekton. But this lack of absolute distinction, which is to be felt in all schemes of classification of natural objects, is no argument against the use of a series of terms which are sufficiently exact, are expressive, and have great practical convenience.

Just because of this practical convenience we also still employ Forbes' regions or zones, or some modification of these. They apply only to the benthos or bottom-living population, but everyone who has dredged or trawled recognises the variation in the characteristic fauna and flora as we descend the sea slope from the upper tide mark. Forbes' littoral zone is the foreshore, that margin of sea-coast between high and low tide marks, the right to which has so often been the subject of vexatious litigation. On the foreshore we find a population, the nature of which varies with that of the sea bottom. The littoral zone of Forbes was a gravelly or rocky seashore and contained abundance of sea-weeds such as the bladderwrack (*Fucus*), the fine green weed (*Enteromorpha*), or the sea-grass (*Zostera*). On it, in our country, are abundance of shellfish like the mussels, dogwhelks (*Purpura*), limpets and periwinkles. The acorn barnacle (*Balanus*) may cover the stones. In the rock pools are crabs, small fishes, worms, zoophytes and many other creatures. Such is the littoral zone of the collector. But we may find that it is a shifting expanse of sand or of sand and mud, and here there are none of the organisms mentioned, or they occur very sparingly and an entirely different fauna and flora are to be found. In their place may be found shellfish like the cockle with other small bivalves, or it may be beds of lugworms (*Arenicola*). In or on the sand and mud are countless millions of diatoms, and these minute plants replace the larger sea-weeds of the rocky shore. Or there may be a gravelly shore covered with a profusion of mussels.

Where, as on the west coast of England, we have rapidly flowing tides with a large rise and fall, and a sandy coast sloping down very gradually into deep water, the foreshore or littoral zone is the region of the inshore fisherman. There, as in Morecambe Bay, where the contours of the "banks" and channels are always shifting, we have the great cockle beds, and there are miles and miles of foreshore densely inhabited by these gregarious molluscs. There too the stake nets, many hundred of yards in length it may be, are set on the margins of the channels. When the sands are covered by the flood tide fishes like plaice, flounders and dabs are caught in these nets and are removed when next the tide lays bare the sands. Here and there are the "baulk-nets," structures of stakes and wattles; and the "hose-nets," long cylindrical nets kept open by rings and furnished with pockets and into which shrimps are carried. When the sands are covered by the tide the smaller boats trawl in the channels for flat-fish or for shrimps, while on the shallower flats, as on the sands near Southport, the trawl, or shank-net rather, may be dragged from a horse and cart, the horse wading up to its belly in the water, with a picturesque, indolent "farmer-fisherman" in the cart. Men wade in the water pushing large nets bent on a semicircular frame of wood and carrying baskets on their backs into which the shrimps caught in the net are put. On the harder parts of the foreshore or on the "scaups," banks, or scars, are mussels, and men go down to these when the tide is out and pick them up. Gradually these mussels beds accumulate mud underneath them until by-and-by a gale demolishes them, rolling up the felted mass of mussels and mud like a carpet, and again laying bare the gravelly bed, which is soon repopulated by young mussels. Hordes of starfishes may invade the mussel beds and may totally decimate them. The echinoderm forces apart the valves of the shell and sucks out the soft body of the mollusc. On the stones of the beach are the periwinkles, which are also picked by hand. On the sands are the lugworm beds which are exploited by the fishermen for bait for themselves and for the seaside visitors. Everywhere on the foreshore except in the most desolate of localities the benthos provides a living, scanty though it may be, for the inshore fishermen.

Forbes' laminarian zone is, as I have said, to be recognised mostly in the areas favoured by the collectors. On such a coast as we have been considering the flat sandy foreshore is continued out to sea below the tide marks for many miles. Here we may have the characteristic sea-weeds, *Laminaria, Himanthalia* and the like, or we may not. There may be a stony bottom with abundant sea-weeds, small rock living fishes and the prawn (*Pandalus*), or there may be a bottom of sand or mud or gravel, or any combination of these, and these latter formations are more important economically, for on them may be immense accumulations of shrimps or of small "flukes" (plaice, dabs, flounders, &c.). This is the rearing ground of many species of fishes such as those I have mentioned and also others such as the dragonets (*Callionymus*), weevers (*Trachinus*), sand-eels (*Ammodytes*), suckers (*Liparis, Agonus* and *Lepadogaster*), gobies, sprats and many others, while codling, whiting, and other young gadoid fishes may live here for a part of their lives. Fish life is very abundant on such sandy shallow waters. I have seen as many as 10,000 fishes taken in one haul of a shrimp net. Hordes of invertebrates also inhabit this region, the commonest of these being starfishes, crabs (*Carcinus, Portunus, Hyas, Stenorhynchus*, hermit crabs and others). Small molluscs like *Mactra, Scrobicularia, Tellina, Hydrobia* and others live in the sand and are preyed upon by the fishes. Zoophytes, sea-weeds, sponges and the larger molluscs like the oyster are generally absent.

Below this again and extending down from 10 to about 50 or 60 fathoms is the coralline zone of Forbes. Ordinary sea-weeds are sparingly present in this region, but we meet with a rich invertebrate population. The nullipores or calcareous algae such as *Lithothamnion* are present sometimes in great abundance, while many species of zoophytes and polyzoa abound. In the shallower parts of this zone (indeed also in the laminarian zone) are the ascidians, the sea-squirts and the beautiful compound forms. The brittle starfishes (*Ophiura* and *Ophioglypha*), the hermit crabs, shellfish like the oyster and scallop (*Pecten*) and the whelks

(*Buccinum* and *Fusus*) are characteristic invertebrates. Here and there are the feather starfishes (*Antedon*), and in the muddy parts about 20 to 40 fathoms the little Norway lobster (*Nephrops*) may abound. The polyzoon *Flustra* (the sea-mat or scented-weed of the fishermen) is one of the commonest invertebrates of the region which we are considering, and with it are to be associated numbers of amphipods. This region is the great trawling ground of the British area. All over it there are fishes like the whiting, cod, gurnard, skate ray, plaice, dab, flounder, sole and many others, while in the deeper parts are the witches, turbot and brill, and in the north the halibut. The hake is found in the deeper water, but this fish is a more conspicuous migrant into our area than the others. The pestilent dogfishes (*Scyllium* and *Acanthias*) are everywhere abundant. All round our coasts in water down to about 50 fathoms in depth fish life is abundant, but below this depth fishes become scarcer.

The deep-sea fauna. Forbes supposed that as we go down into the deep water beyond the laminarian zone life gradually became less and less abundant until at a certain depth, which he fixed in the Mediterranean at about 300 fathoms, it became extinct. It is true that the shallow waters round the coasts contain the most luxuriant life. Rivers here bring down fresh water carrying in solution the food salts for the support of plant life, and because of this and the complete lighting of the whole thickness of the water layers plant life, both the larger algae and the minute diatoms, are most abundant here. The tidal streams, which are stronger near the coast, distribute this ultimate food material brought down by the rivers and carry it along the littoral waters. Life here is thus relatively abundant and as we pass out into deep water it certainly becomes less so, but to say that at any depth it is altogether absent is quite erroneous. Even to say that life is very scanty in the great oceanic abysses may not be justifiable, for it is an operation involving much expense of time to dredge or trawl in very deep water, and it is certain that but an insignificant portion of the sea bed at the great depths has been explored. The shallow parts of the sea are very well known, at least with respect to their benthic fauna and flora. It is hardly an exaggerated statement to make when one says that there is not a square foot of the North Sea which has not been sounded by the fishermen, and certainly there is not any part of that sea where it is possible to trawl that has not been swept by the net. When one compares our knowledge of the shallow seas with that of the deep ocean bed, derived from the very few hauls of the trawl or dredge, which have been made there it is not so certain that our experience justifies us in assuming that animal life there is enormously less abundant than in the inshore seas. Certainly the deep sea is not even relatively sterile.

Before the time of the great oceanographical voyages naturalists thought of the exceptional conditions prevailing at the bottom of the deep oceans: the utter darkness, compared with which even the obscurity of a moonless night is relatively light; the enormous pressure which organisms living there have to undergo (in 3000 fathoms approximately three tons to the square inch); the uniformly low temperature which they correctly supposed to obtain in the depths; the supposed stagnation or imperfect aeration of the water at the bottom—all these conditions pointed to the absence of life, and such *à priori* speculations could only lead to one conclusion, that the great oceanic abysses were regions of desolation.

When isolated observations made before the voyage of the *Challenger* rendered it probable that the deepest recesses of the ocean were the abode of life, expectations of strange results when these were adequately explored were entertained. It was probable that the geological antiquity of the ocean beds was very great, and since conditions must be very uniform there it was imagined that deep-sea trawling would reveal "living fossils," animals akin to those which flourished during past geological epochs. Now while many characteristic deep-sea animals present archaic features—the stalked crinoids for instance—it is nevertheless the case that these expectations have been disappointed. The deep-sea fauna is peculiar, but all the same it exhibits a general resemblance to that of shallow water. There is no deep-sea flora for, since light is entirely absent except for that which is afforded by the phosphorescence of most of the animals residing there, plants cannot exist. But with this exception all the groups of living things found on shallow water are represented. Echinoderms such as the brittle stars, and the stalked feather stars which are the "last survivors of a large and important order which flourished in past geological ages," occur, and the sea cucumbers or Holothurians are also found. Sponges form an important part of the deep-sea fauna. Corals, zoophytes and their allies, though not so numerous as the sponges, are nevertheless well represented. Molluscs and crustacea in all their classes are found. Worms are found in the red clay of the deepest abysses. Fishes are uniformly and relatively abundantly distributed. Generally speaking all the shallow water groups have their deep water allies, but in the depths there are strange adaptations and bizarre forms. The variety of animal groups present, the numbers of species, and the curious contrivances for obtaining food all point, it has been said, to the conclusion that in the deep sea the struggle for existence is no less severe than in the shallow waters near the land. Possibly then the abundance of life in the depths is great and there is an insufficiency of food. This condition would produce a struggle for existence.

Plants are, as we shall see, the "producers of the sea." They alone (with some of those "borderland" creatures which we have difficulty in classing) can form organic substance out of inorganic material. If then plants are absent at those comparatively shallow depths at which light fails to penetrate the sea how can the fairly abundant deep-sea population obtain its food? Obviously these animals cannot live wholly on each other like the inhabitants of the Scottish village who earned a living by taking in each others' washing! Food must be conveyed to "denizens of the deep" from some outside source, and as a matter of fact we find that the majority of the abyssal animals feed on the bottom deposits. We have seen that with the exception of perhaps the red clay much the greater part of the deep-sea deposits is formed from organic remains. All the organisms which make up the organic oozes live in the upper productive layers of the sea and when they die their bodies fall to the bottom. At the low temperature found there bacterial activity is largely inhibited and therefore putrefactive processes are carried on very slowly, so that for a considerable time after they reach the sea floor planktonic organisms, like diatoms or protozoa, must retain a considerable proportion of matter capable of serving as food for the deep-sea benthos. The abyssal deposits therefore serve as a kind of "pap" which is eaten by the bottom-living animals for the sake of the nutriment it contains in much the same kind of way as a lugworm eats the sand, or rather passes this through its intestine and digests and absorbs the small proportion of organic substance which the sand contains. These ooze-eating animals are then the prey of others, and since the deep-sea fauna is to a great extent a predatory one we find that many of the abyssal crustacea are provided with queer organs of attack and defence: tactile, prehensile and alluring contrivances, all used in the effort to obtain food or in that of resisting the attacks of their enemies. Some crustacea for instance are of enormous size when compared with their shallow water allies, and many deep-sea fishes have enormous heads furnished with great mouths armed with formidable teeth, while their stomachs are large and the bodies and tails are small and fragile. Although daylight never penetrates to the bottom of the deep sea it is still the case that eternal darkness does not prevail, for many of the abyssal creatures are phosphorescent. Fishes may have rows of luminous organs, and some have a kind of bull's-eye lantern from which a stream of light can apparently be thrown at the will of the animal. Probably a kind of diffuse light is present in the deep, and at any rate animals may be visible by means of their own phosphorescence as they move about. Some are blind but others have large eyes. Associated with the phosphorescence are the prevailing colours of deep-sea animals. Glaring reds and browns prevail and generally monotones are the rule, and the diversified colours of the upper world are absent. Curious problems arise in the consideration of the mode of life of these creatures. How is it, for instance, that they can withstand the incredible pressure of the superincumbent layers of water? This is a question which has never been properly answered. Experiments with ordinary marine animals subjected to enormous pressures seem to shew that profound bodily disturbance results from this. The first effect of increasing pressure is the excitation of the nervous system and then follows the inhibition of the functions of this. The animal is thrown into a state of coma from which it recovers if the pressure is released and if it does not continue too long. These effects are produced by the inhibition of water by the tissues[1]. In what way the abyssal animals have become adapted to the great pressures which they undergo without apparent inconvenience is difficult to understand. The effects produced on a deep-sea fish when it is suddenly brought up from its habitat are well known. The intestine or stomach is usually projecting from the mouth and the flesh and bones are disintegrated by the release of pressure and the consequent expansion of the liquid or gas contained in the tissues.

In the depths of the ocean the temperature is always uniform, and is nearly the same over vast areas. Over great extents of sea floor the bottom deposit is the same. Currents are very feeble. The composition of the water does not vary. We find then that in accordance with this sameness there is great uniformity in the distribution of the fauna. Some deep-sea species are cosmopolitan and many are very widely and uniformly distributed.

The migratory fauna. So much for the benthos or bottom living animals and plants of the sea. We have seen that its constituents are the larger sea-weeds and the invertebrates. The nekton or actively locomotory organisms in the sea are almost entirely the vertebrates. These are they which can move about in obedience to their volitions, or in response to instincts, or as the reflex to physical changes. They migrate independently or in opposition to the currents or drifts which carry about the feebler planktonic species. The marine mammalia are the chief among the nektic animals, and these, with the great sharks, roam over extensive tracts of ocean and probably make cruises which are comparable with those of the large ocean tramp steamers. The fishes are the type of the nekton. Some, like the flat-fishes, are often spoken of as semi-sedentary, but it is more probable that even these are nearly always "on the move." Others like the cod, hake, herring, salmon and many others are notorious wanderers and swim over wide areas of sea. The fresh-water eel it is now known makes an extraordinary long spawning migration, and, as Schmidt has shewn[2], the eels of the north of Europe migrate

---

[1] Regnard, *La Vie dans les Eaux*, Paris, 1891.
[2] Schmidt, *Rapports et Procès-Verbaux; Cons. Perm. Int. Exp. de la Mer*, Vol. v. Copenhagen, 1906.

through the Baltic or North Sea or both, or through the English Channel in order to reach their spawning ground in the North Atlantic south-west of the Faeroe Islands, while the comparatively feeble larva or Leptocephalus makes a correspondingly long journey to reach the rivers into which it ascends to pass through its adolescent phase. The migrations of the sturgeons occasionally caught in the English Channel or in the Irish Sea may be compared with those of the migratory birds. Some invertebrates are also nektic animals, as for instance the lobsters, crabs, and some other crustacea, which are powerful and intelligent animals and move about as they please. Cuttlefishes and squids certainly do so. Some worms may belong temporarily at least to the nekton, and the large medusae, though perhaps better classed with the plankton, do move about " of their own accord." Like the benthos of the moderately shallow seas the nekton varies of course with the locality.

The plankton. When we have considered the organisms of the benthos and nekton we have apparently exhausted the visible life of the sea. But by far the greatest proportion of this must belong to the plankton. The mariner, it has been said, when he sails over a tract of sea thinks that he traverses a "barren waste of waters" through which there swims here and there an occasional fish or porpoise. But in reality he sails over a "pasture," and beneath his ship is a wealth of life much more abundant than is contained in the richest or most luxuriant forest. Beneath his feet may be a couple of miles of water and every cupful of this may teem with life, and this is so even if no fishes or other large animals may be visible. This enormously abundant life is the plankton, the drifting fauna and flora of the sea, the presence of which is only revealed by the tow-net and microscope. Of all forms of marine biological investigation the study of the plankton is the most entertaining. Equipped with the tow-net and microscope the naturalist finds here a veritable "wonderland" awaiting him, and the variety and beauty of the creatures so obtained, and the ever present possibilities of finding forms of life new to science, combine to make the study of the plankton a most fascinating one. To any one who lives near the sea the observation of the microscopic life may be a continual occupation and delight. But in addition no department of biology presents more abstruse problems to those who care for such investigation.

All the great animal groups with the exception of the higher vertebrates are represented in the plankton. It will be convenient to consider here, in a general sense of course, what organisms are present, premising that there is a very considerable variation both with the locality and with the time of year when the plankton is observed. In the floating microscopic life of the sea there may be distinguished two categories of organisms, (1) the transitory plankton, that is the assemblage of forms which appear only for a short time and then assume other habitats: such are the larvae and young forms of many of the animals and plants living among the benthos and nekton, obviously the place and time of appearance of these will vary with the character of the bottom fauna and that of the nekton. Then we have (2) the permanent plankton which consists of those organisms which live for the whole term of their lives in the sea as drifting creatures, and these too vary from place to place just as the other forms of marine life do.

Fishes appear in the plankton in the form of eggs and larvae. Though, properly speaking, they are nektic animals, they nevertheless make a very obvious part of the plankton in their young stages and some (the skates, rays and dogfishes) belong, while they are developing, to the benthos. All the British skates and rays, some of the dogfishes, and some other fishes, of which the herring is the most important, lay eggs which, being heavier than sea water, sink to the sea bottom, and after remaining there for a certain period of time hatch out. The herring eggs undergo a period of incubation of about a fortnight, at the end of which time the young fish swims out into the water and takes its place among the plankton. The eggs of the skates and most of the dogfishes are large and are provided with a strong capsule or shell. The period of incubation lasts for the greater part of a year and at the end of this time the little fish hatches out and at once settles down to its natural habitat on the sea floor. The empty capsule—the "mermaid's purse" of the children—is by-and-by washed up on the beach by the tides and if one opens it at the right end (the right end is difficult to find) he will find a guinea inside! Some of the dog-fishes (*Acanthias*, the spur-dog, is the best known) are oviparous and the young are born alive, and on issuing into the sea at once form part of the nekton.

But most of the other fishes of the North-Atlantic ocean produce eggs which are pelagic, that is they are lighter than sea water, and when they are shed by the parent they rise to the surface, or near to the surface, and there they drift about until the embryo hatches out from the shell. These have absolutely no powers of locomotion and they are drifted about passively by tides and currents, the very type of planktonic organisms. After a period of incubation which lasts for a variable time—one week to three—the larva emerges. It is at this stage one of the feeblest of marine creatures, and though when seen in the watch-glass of the planktologist it is a large and active creature, so far as microscopic animals go, it is carried about wherever winds and currents drift it. The superficial water drift may transport it to regions remote from that in which it was spawned, and countless millions must be borne into the brackish water at the mouths of rivers, where they probably perish, or they may be stranded on the shore. But after a month or so the little fish has grown considerably, and now ensues the period of metamorphosis, during which the larva takes on the adult characters and, in the case of most edible fishes, sinks to the bottom of the sea to pass the rest of its life, or, if its parent is a pelagic fish such as the herring or mackerel, it begins its career as a member of the pelagic nekton.

Fishes are the only representatives of the true vertebrates which appear in the plankton. But there are still the ubiquitous Ascidians, animals which hang on to the vertebrated skirts as degenerate and poor relations. Such are the sea-squirts, and the brilliantly coloured compound ascidians which we find on the weeds and stones of the beach or dredge up from deeper water. The more primitive tadpole-like Appendicularians—"tails which wag heads"—are nearly always present in the plankton, each inhabiting its gelatinous house (*Haus*). These are permanent members of the drifting category of organisms, but the eggs of the other ascidians form very frequently occurring objects in the plankton. Fishes and ascidians—chordates in strict zoological

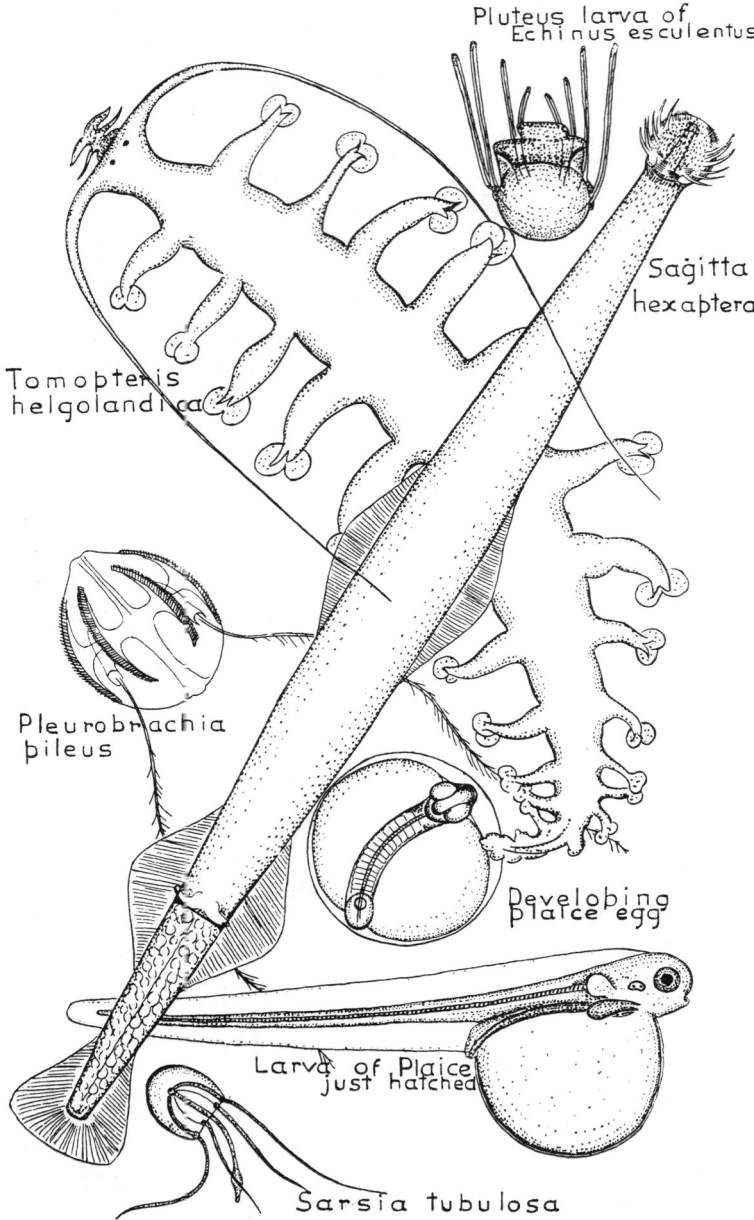

Fig. 18. Plankton: fishes, worms, &c. Magnified.

terminology, are important, though not relatively abundant constituents of this category of marine life.

The **Crustacea** form one of the great groups of the plankton, just as the same animals form perhaps the predominant inshore benthic population. We may attribute to the crustacea the same predominance in the life of the sea that the insects possess on the land. Haeckel[1] has pointed out that the struggle for existence has called forth the greatest variety in form and the most wonderful adaptations in structure, habits and life-history among the insects; and that just so the no less severe struggle in the sea has produced an even more wonderful diversity in form and habit among the crustacea. It is among this group of animals that we find some of the most intelligent and powerful in the sea; the great lobsters and crabs for instance, which exhibit a high degree of specialisation in their structure and in their habits. Here too we encounter in the parasitic crustacea some of the most degenerate and degraded of animals. The crab-barnacle (*Sacculina*), for instance, is by reason of its weird life-history one of the most wonderful animals in existence. Crustacea are always abundant and are quite universally distributed in the sea. Though they do not possess the astonishing fertility of the fishes nevertheless they exhibit a greater amount of specialisation in the varieties of devices for the care of offspring which makes for the greater fecundity of the class.

To the benthos belong the lowest group of the crustacea, the Cirripedes or barnacles, and the highest, the crabs, prawns and their allies. All these produce pelagic larvae which take a very important place in the plankton. The barnacles (*Balanus*) which form so abundant a part of the littoral fauna produce incredibly great swarms of larvae (Nauplii) which sometimes appear in the plankton to the exclusion of everything else, and for a time these nauplii drift about and then undergo metamorphosis into the Cypris stage and settle down on almost any object in the sea as the familiar little acorn-shells. So also with the crabs and their relations, the prawns and lobsters: these in their adult stages are comparatively large animals and produce great numbers of eggs.

[1] In the *Plankton-Studien*.

The larvae which hatch out from these eggs are the Zoeas and these judged by microscopic standards are fairly big creatures. Zoeas have a comparatively long life in the sea as pelagic animals before they undergo their first metamorphosis into the Megalopa stage, and the megalopa has also a somewhat long life in this stage. By-and-by the megalopa, by repeated moults, passes into the adult stage and the Decapods then settle down to the sea bottom for the remainder of their natural lives. At times the tow-nets may contain nothing else than the zoea or megalopa stages of crabs. Just so the shrimps and prawns appear also in the plankton in their larval stages. Other groups of crustacea, the Isopods, Amphipods, Phyllopods and Stomatopods, terms which the unprofessional reader will understand as connoting groups of crustacea distinguished by well-marked variations in structure, also occur in the plankton, at times in quantity. Some are permanent planktonic animals, while others live on the sea bottom as adults and only appear among the pelagic life of the sea in their larval stages; but many others, many of the Schizopods and Ostracods, are permanent inhabitants of the free ocean and shallow seas.

The crustacean animals known as the Copepods are by far the most abundant metazoan animals living in the sea. Some of the single celled animals or protozoa may surpass them in number, but the importance of the copepoda in the economy of the sea depends not only on their number but also on their size, and many thousands of some of the smaller protozoa may not have the bulk of a single large copepod. The copepoda are small creatures, the average length of which in our seas may be put at about one-sixteenth of an inch. They are quite ubiquitous. It is so rarely that a tow-netting can be taken without including some of these micro-crustacea that when this occurs the planktologist feels that the occurrence is worthy of special remark. The copepods are for the most part pelagic in habit, but a very considerable group of them is found in the mud and sand at the sea bottom and some have adopted a parasitic habit and live on the outsides of many fishes. The pelagic forms have as the figure shews, a somewhat cylindrical body, ten pairs of appendages, two pairs being feelers, and a jointed tail furnished with hairs (the importance of

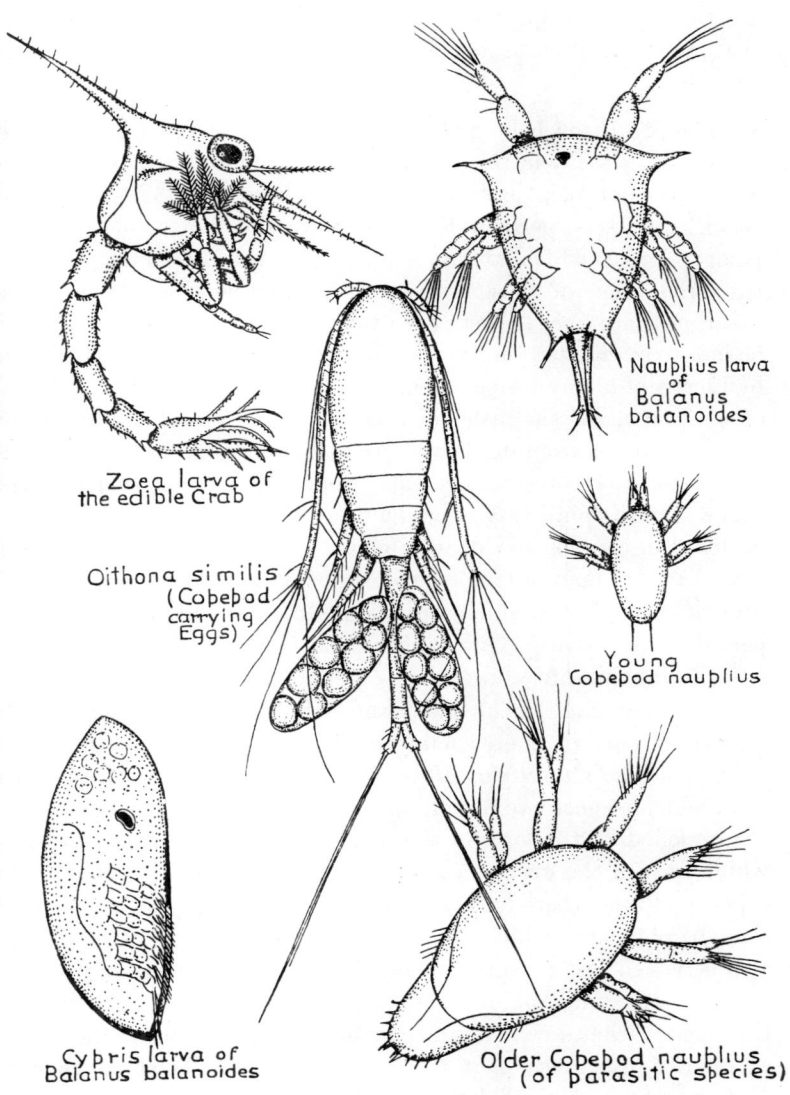

Fig. 19. Plankton: Crustacea. Magnified.

these animals impels me to this technical description). There are thousands of species of copepods in the sea and it is certain that there are hosts of new forms which have not yet been described.

The **Mollusca** belong characteristically to the benthos and only appear transitorily in the plankton. Many molluscs produce larvae which are not pelagic, but there are others, such as the hosts of cockles, mussels, periwinkles, and other shellfish, which produce pelagic larvae which at times appear in the tow-nets. It is by the production of these free-swimming larvae that the molluscs mentioned are able to distribute themselves, just as animals which have a fixed habitat on the sea bottom are generally distributed by means of larvae which are carried about in the sea. A "strike" or settlement of shellfish such as the mussel and cockle occurs when the free-swimming larvae are abandoning their pelagic mode of life for that on the bottom. Some of the gasteropods (the shellfish with spiral shells) produce eggs which in many cases are enclosed in cocoons and do not float in the sea, but in this group also there are many which have pelagic larvae. There are also two groups of Mollusca which are planktonic animals for the whole period of their lives. One of these, the Heteropods, are characteristically inhabitants of warmer seas, but the other, the Pteropods, are most abundant in the polar and subpolar regions, and play an important part in the economy of the sea. These creatures, the winged shells (*Farfalle di Mare*), are usually enclosed in thin shells from which project two fins or vanes by means of which the animals are impelled by the force of the currents. It is these latter animals which furnish the greater part of the food of the whales. Of the other mollusca there remain two small groups, the Chitons and the Scaphopods, which are benthic, and the cuttlefishes and squids, which are pelagic. These latter are the most highly developed and formidable invertebrates in the sea. Indeed it is only the sperm whales which are capable of dealing with the giant calamaries or squids. But they are active animals which, though in some cases pelagic, belong rather to the nekton.

Among the heterogeneous groups of **Worms** ("the cross of suffering for systematic zoology") there are only a few groups which are planktonic animals throughout their lives. Conspicuous among these are the arrow worms (*Sagitta, Spadella*, &c.) which are ubiquitous and sometimes appear in the plankton to the exclusion of all other kinds of organisms. Some of the true worms (*Chaetopoda*) are pelagic dwellers, as for instance the beautiful creatures *Chaetopterus* and *Tomopteris*—and these in some seas are abundant. Then we have many worms which produce pelagic larvae, which for a time are conspicuous members of the plankton but are of little importance in the general economy of the sea.

The **Echinoderms** (the starfishes, sea-urchins and sea-cucumbers) are a numerous and varied marine "phylum." They are exclusively marine and none is an inhabitant of fresh water. All live at the sea bottom and only one or two, such as the Comatulidae, and sometimes the common starfish (which occasionally swims in the manner of a medusa by the rhythmical contractions of its arms), appear exceptionally in the nekton. But the larvae of some of the sea-urchins (as for instance *Echinocardium cordatum*) are among the most beautiful of the forms which can be taken by the tow-net. Many plutei and other larvae of echinoderms may be found among the transitory plankton, but none is of importance in the metabolism of the sea.

The **Coelenterates** (jelly-fishes, zoophytes, corals, &c.) are very important planktonic groups. They are animals in which there is no distinct body cavity, that is the body wall is a structure which consists of two layers only and encloses a single cavity which serves as a stomach, alimentary canal and body cavity, and has a single opening which is surrounded by a circlet of tentacles as in the familiar case of the common sea-anemones. The latter coelenterates are solitary animals, but the majority of the class live as colonies as in the case of the zoophytes, in which we have a number of individuals living in association connected together by a common flesh and all growing on a plant-like skeleton; and the corals which also live as colonies but which form massive limy skeletons which build considerable reefs and other formations. In the case of these colonial animals the group which is the unit, is formed by the budding of an originally single "zoid," though sexual reproduction of course occurs among the coelenterates. These animals produce larvae which make an evanescent appearance in the plankton, though they are, in their final phases, benthic organisms. Two great groups of coelenterates are permanent and important planktonic animals. The medusae are very conspicuous pelagic organisms. Every one who has been to sea must have seen the immense shoals of jelly-fishes which at times cover the surface of the sea. I have sailed for several miles through a shoal of *Aurelia*, in which the separate animals were closely packed and were piled on the top of each other completely discolouring the sea. Then there are the large white "Cabbage-blebs" (*Rhizostoma*) which are at times so numerous as to form a great nuisance to the fishermen. The medusae swim by the contraction of their bells, but their powers of movement are not great enough perhaps to justify us in including them with the true nektic animals. Another group of coelenterates, which are of great interest by reason of the extreme beauty of some of their species, are the Siphonophores, of which the best known example is the Portuguese man-of-war. They are inhabitants of oceanic areas and generally of warm water.

The **Ctenophores** are very common planktonic organisms. In our seas the "Marble-bleb" (*Pleurobrachia*) is the only very common species. This beautiful little creature must be familiar to any one who has seen a shrimp trawl hauled in our seas in the summer. It is very often present in the tow-nets though not in very great numbers, as its size prevents this apparatus from catching it in abundance. In warmer seas the pelagic ctenophores are beautiful and conspicuous objects of the plankton.

The **Protista** are the most abundant of the planktonic organisms and by far the most difficult to study. They are for the most part animals of very small size and on this account they are not represented in the catches of plankton made in the ordinary tow-nets to the extent that they occur in the sea. No silk net has yet been constructed which will take a representative sample of these organisms from the sea. The protista are unicellular plants and animals, or they are those anomalous forms which fit with difficulty into any wide scheme of classification. We can divide them into the Protozoa, which are single-celled animals, and into the Protophyta, which are single-celled plants. But this division is far from being an absolute one and there are many forms which may be placed in either category.

The Infusoria are Protozoa which play a more important part in fresh-water than in marine plankton. Most of them are littoral organisms living in the mud or on the surface of water weeds. Only two groups of infusoria are of importance in the sea. The Flagellates are represented by *Noctiluca*, a protozoan which occurs at times in the sea to such an extent that the water may be, to the naked eye, visibly discoloured by it. It is the principal cause of the phosphorescence of the sea in some regions. Its distribution is sometimes peculiar; thus it occurs on the west coast of England and Wales in enormous shoals, which have a very wide distribution within the limits mentioned and occur during the greater part of the summer and autumn. At times it occurs in the inshore waters, even in the estuary of the Mersey up beyond the Liverpool landing stage in such masses that the sea may be brilliantly phosphorescent and a tow-net dragged through it may contain this organism to the exclusion of apparently everything else.

The Peridinians, of which *Ceratium* is the most common genus, are shelled infusoria provided with flagella. *Ceratium* is, like *Noctiluca*, also a very common cause of phosphorescence of the sea. Allied to the Peridinians are the Ciliata, of which the Tintinnoidae are the most common planktonic representatives. They too are shelled animals, but in their manner of nutrition are rather to be regarded as approaching to the unicellular plants. Many of them are extremely elegant in form. Some are neritic, or inshore forms, but many are oceanic in their habitat.

The best known groups of the protozoa which occur in the plankton are the Foraminifera and the Radiolaria. It is the shells of these extraordinarily numerous animals which form the wonderful deep-sea deposits, the Globigerina and Radiolarian oozes. It was formerly thought that those organisms, the shells of which made up the Globigerina ooze, belonged to the benthos, but the discoveries of the *Challenger* proved beyond doubt that the Foraminifera making up this deposit are animals which when alive live in the open sea at a great variety of depths.

They are distinctively oceanic organisms. Thus the *Challenger* found "great banks" of pelagic foraminifera in the equatorial Pacific, and indeed they occur everywhere in the open sea, particularly in warmer waters in "numberless myriads." When

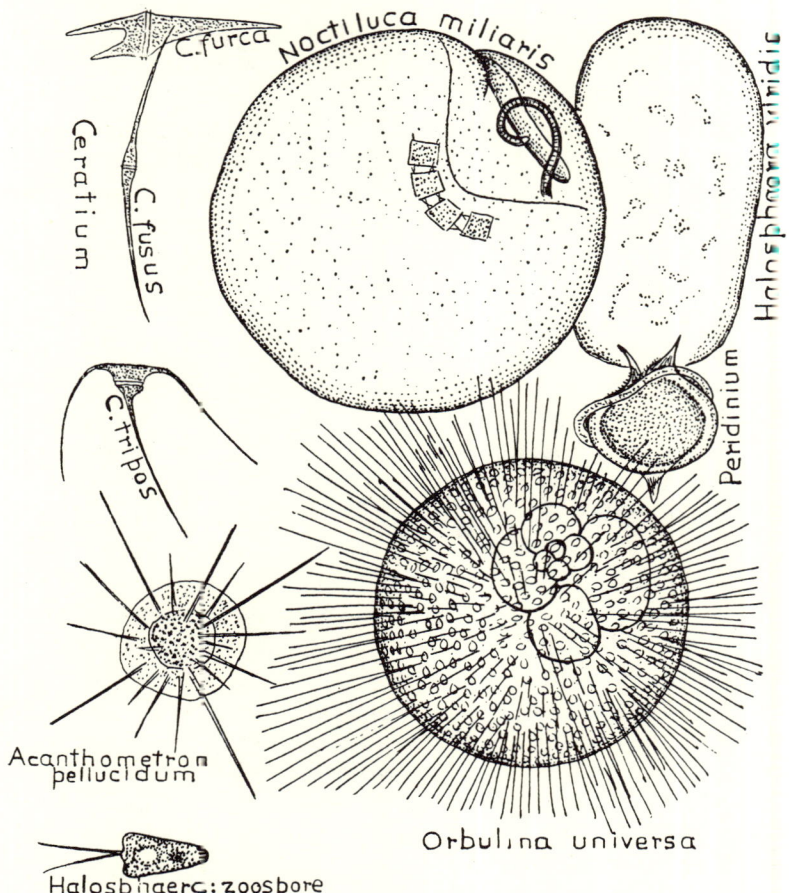

Fig. 20. Plankton: Protozoa, &c. Magnified.

obtained from the deep-sea oozes foraminifera only shew a rounded shell (the appearance of the well-known *Globigerina bulloides* is now familiar to every one) but when seen in the living state the organism is a thing of great beauty. The shell is sometimes furnished with long spines and is pierced with holes through which the protoplasmic body of the animal is protruded. Foraminifera occur, of course, in the inshore waters but they are distinctively oceanic organisms.

The Radiolaria are also oceanic planktonic protozoa which, like the foraminifera, are provided with skeletons; but the latter, which in the group already discussed are composed of lime, consist of silica. The form of these shells is extremely elegant. Radiolaria are common in seas which contain the normal amount of salt, and therefore are not abundant in inshore waters. In the colder seas they may occur in great numbers, but the greatest variety of species is to be found in the warmer oceans. They occur at all levels in the sea and when they die their siliceous skeletons sink to the bottom and form the well-known Radiolarian ooze. The "fabulous radiolarian treasures" brought home by the *Challenger* from the bottom of the seas of the globe when worked up by Haeckel led to the discrimination of 4318 different species, and it is quite certain that many more forms of radiolaria remain to be described[1].

The **Diatomaceae** are above all the most important organisms in the sea regarded from the point of view of their significance as the producers of organic substance. The diatoms are the "pastures of the sea" and correspond to the "grass of the fields" of the land. They are quite ubiquitous and are found alike in fresh water, in the sea, at the surface and at the bottom of shallow waters, in mud, sand or on the surface of weeds, &c. They are true plants and are accordingly found only in those parts of the sea into which sunlight penetrates, though the amount of the latter which is sufficient for their metabolism is far less than sunlight in its full intensity. Like the foraminifera and radiolaria they occur in the sea in such incredible numbers as to form immense deep-sea deposits by the accumulation of their dead shells. Diatoms are sometimes called "unicellular Algae." They are vegetable cells containing a nucleus and a variously shaped

[1] Haeckel described 4318 species of Radiolaria. These belonged to 739 genera, which again were grouped into 85 families, and these families were derived from 20 orders, while the latter again could be derived from one " common ancestral form," *Actissa*. It seems "too good to be true." (See *Report on Challenger Radiolaria*, 1887.)

chromatophore, the structure containing the chlorophyll essential for the metabolism of the organism. This cell is enclosed in a siliceous shell consisting of two valves which fit over each other in much the same manner as the lid of a pill-box fits on to the box. Reproduction is carried out by the division of the cell within the shell. After the cell divides two new valves are formed, one valve within each of the two which formed the original shell of the diatom. It follows from this that at each division the daughter diatoms must be smaller than the mother cell, and further that this process cannot continue indefinitely. This is the case as a matter of fact. After a time the process of reproduction by division is superseded by a process, in the course of which two diatoms of the same species come together and become surrounded by a gelatinous investment. The contents of the two shells, that is the diatom cells, then issue from the shells and mingle together and a structure known as an auxospore is formed and this becomes a diatom of the typical form in which the process of division is again set up. When circumstances become unsuitable, as when the water in which the diatom is contained dries up, the cell can form a resting spore, that is the protoplasmic contents become gathered up into a compact mass which is able to resist unfavourable conditions.

An almost infinite variety of shape is exhibited by the shells of diatoms. We find that those resident in fresh water have forms which are quite typical of this habitat and differ notably from those of the marine species. *Navicula*, for instance, is a type of the bottom living diatoms both of the sea- and of fresh-water. In both the fresh-water and sea-water flora we have the peculiar stalked forms *Gomphonema* and others. In the mud are the littoral genera *Navicula* and *Pleurosigma* with an incredible number of species. Both of these forms are diatoms shaped somewhat in the fashion of little double-stemmed boats. They can move about in a graceful manner and by means of a mechanism the precise nature of which the botanists do not yet seem able to determine, in spite of the enormous amount of work which has been done on these organisms both by amateur and professional workers. In the sea the principal pelagic forms are *Coscinodiscus*, a diatom which is shaped like a pill-box, and some species of which are relatively large; *Biddulphia*, a diatom the typical form of which

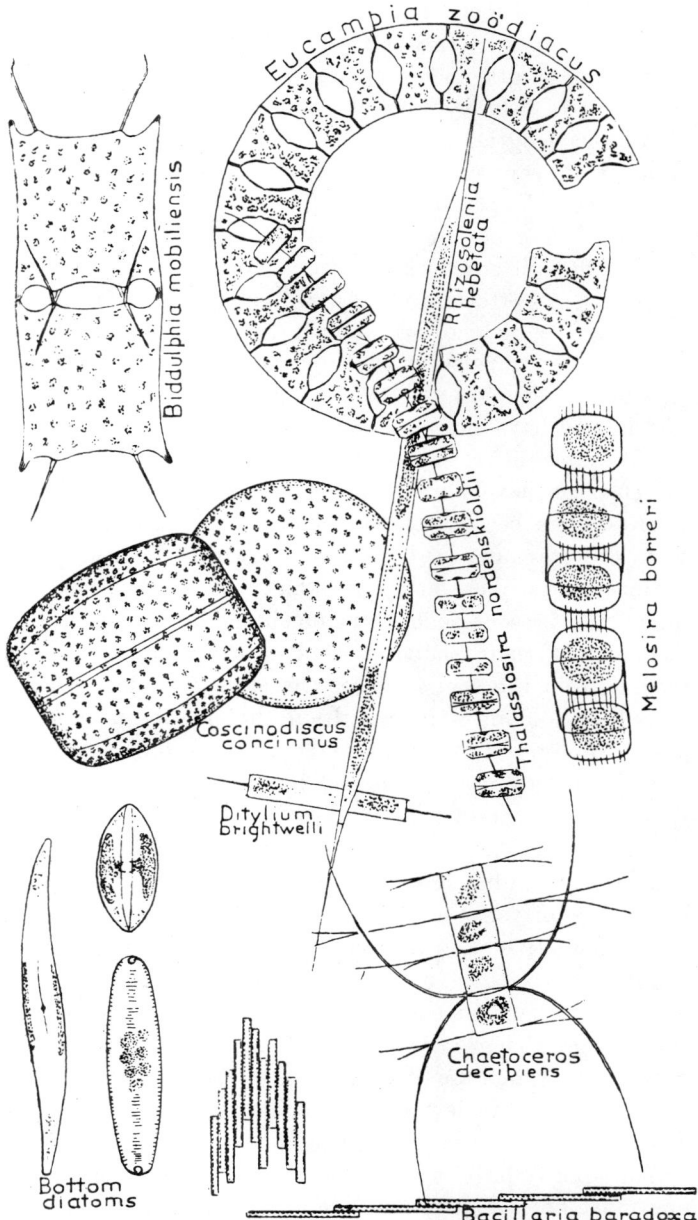

Fig. 21. Plankton: diatoms. Magnified.

is in shape something like a pillow provided with variously arranged spines at the angles; *Rhizosolenia*, a cylindrical diatom of varying length and dimensions; *Chaetoceros*, in which we have a number of short cylindrical units adherent together with spines of various lengths and manners of insertion; *Thalassiosira*, in a number of separate units are strung together on a filament; *Bellerochea, Nitzschia, Bacillaria, Asterionella* and many others. Some of these are neritic forms but others are oceanic dwellers. In the diatom ooze, which covers more than ten million square miles of sea bottom in the Antarctic, we have a variety of forms the chief of which are species of *Fragillaria, Coscinodiscus, Thalassiothrix, Navicula, Actinocyclus*, &c. It will be seen then that many genera of diatoms are cosmopolitan in their distribution, and indeed have the most varied habitats. The species of these are of course different in widely different localities.

Diatoms have always been a favourite study with amateur microscopists because of the ubiquity of their distribution both in fresh waters and in the sea. No form of microscopical investigation can be more attractive than that of the diatoms. These organisms present such an astonishing variety in form, and the elegance of their shape as well as the extreme beauty of the markings on the shells allures the student. But in spite of the incredible amount of investigation that has been lavished on the speciography of this group, much remains to be discovered with regard to the reproduction and life processes. We will see later that the study of the diatoms has been taken up in relation to far-reaching questions of the general economy of life in the sea, and that the stimulus afforded by the necessity for more minute knowledge of the metabolism of these organisms can hardly fail to be productive of a great increase in our knowledge of the group.

This is, of course, only a very brief sketch of the composition of the plankton, and I have not referred to several very interesting sub-groups. There are hosts of planktonic protozoa like the Flagellates, which are of very great interest, both biologically, and from the point of view of the circulation of food matter in the sea. Then there are the anomalous forms, like the Challengerida, which afford quite special problems in speciography. Finally one must not forget to mention that many algae (like *Trichodesmium*) are planktonic.

CHAPTER IV.

LIFE IN THE SEA (*continued*).

WE find therefore that there are two main groups of organisms in the plankton of the sea. There are first of all those which are pelagic throughout their lives: these are the permanent plankton. Then we have very many animals which appear only for a short time in the plankton and then forsake this mode of life and adopt either a benthic or a nektic habitat. To the permanent plankton belong the diatoms and the other drifting algae, almost all the protozoa, many of the jelly-fishes, many worms, some molluscs and several very important groups of crustacea. On the other hand there are also very many organisms which form what we may term the transitory plankton: all these are eggs, larvae or other developing stages of marine organisms, and among them we have the eggs and larvae of fishes, molluscs, crustacea, echinoderms, worms, with various stages in the development of most zoophytes and actinians, and also the spores of those algae which are in their adult state fixed organisms. For a certain time in the year all these appear as drifting pelagic organisms and then they settle down on the sea bottom, or adopt a definitive nektic mode of life.

This interchange between the plankton, nekton and benthos depends on the fact that an animal or plant, which for the greater part of its life is rooted to one spot, must possess some means whereby its eggs may be distributed over a much wider area than is occupied by the adult forms. Thus on the land the seeds of plants are carried about by winds and other means, so we find also in the sea that all benthic and many nektic animals produce eggs which are carried by the currents over a very wide area of sea. In this way the species becomes distributed over as wide an extent of sea as is consistent with the climatic conditions under which it is fitted to live. Now it is the existence of the method of reproduction by the production of free-swimming larvae which is the cause of the transitory appearance of a whole host of different classes of marine life among the plankton.

Marine animals exhibit various methods of development. The two highest groups of vertebrates in the sea—the mammalia (the seals, whales, &c.) and the reptiles (the turtles and water-snakes)—reproduce in the same way as their relations on the land, that is the reptiles lay large eggs which develop in the same way as those of birds, while the mammalia are viviparous. Among the fishes there are some which reproduce in a manner which is very similar to that found among the mammalia; others lay large eggs which develop in a manner not unlike that seen in the eggs of the birds and reptiles; and others again lay eggs which develop in a very short time into a larva.

In the development of a warm-blooded animal the embryo grows within the uterus of the mother and is nourished by a placenta, that is a structure in which the nutritive matter of the maternal blood diffuses through into the blood of the foetus. This kind of development occurs exceptionally among the fishes. There are a few British forms, and many tropical ones, in which a viviparous mode of reproduction obtains and the little fish is born from the body of the mother just as a mammal is. Not only so but in some fishes there is even a structure which resembles the placenta of the mammalia. Thus in some tropical rays there is a functional placenta. The yolk sac comes into relation with the internal wall of the uterus, and little projections grow out from the latter and come to fit into depressions in the wall of the yolk sac, and the nutritive matter required by the growing embryo diffuses through the walls of these blood-vessels. Then in the British spurdog (*Acanthias*) we have an example of a viviparous fish. If a gravid female caught during the summer be opened four or more young dogfishes will be found in the uteri and if the size and weight of these be compared with those of the undeveloped eggs it will be seen that a notable increase in both weight and size has occurred during development. The blood of the mother must therefore supply some material to the embryo other than that provided in the egg. Other British dogfishes and rays lay large eggs containing a quantity of food yolk, thus the eggs of the spotted-dog (*Scyllium*) and those of the skates and rays are fairly large and are enclosed in a leathery looking capsule, and are deposited on the sea bottom. Development takes a long time, perhaps the greater part of a year, and while the embryo is growing inside the egg-capsule it is being nourished by the food yolk originally present in the egg. In fact the general course of development is very like that exhibited by the egg of the fowl, except that the latter develops in a much shorter time, because it is incubated at a temperature of about 38° C. while the egg of the skate or ray is exposed to a temperature which does not usually rise above 15° C. In other respects the reproduction of these fishes is curiously similar to that of the fowl, for the fish probably "lays" in much the same manner and the spawning period probably extends over a considerable part of the year. At the end of the incubation period the little fish which is hatched from the egg is a large and perfectly formed animal and one which has probably no difficulty in escaping from its enemies, or in finding its food.

But the eggs of the other fishes are quite different from those of the skates, rays and dogfishes. Most bony fishes produce a great number of ova, and these are very small, usually one to two millimetres in diameter, and they are therefore provided with a very small quantity of food yolk. Some, like the eggs of the herring, are laid on the sea bottom (demersal eggs), but most fishes shed their ova into the sea and the latter then develop floating about among the plankton: these are pelagic eggs. On account of the limited amount of food yolk development is a rapid process and the little fish usually hatches out from the egg in a week or two, but is a very feeble and helpless creature. It is quite unable to feed for itself, indeed the gullet is usually not an open tube for some time after it is hatched. The larva, for such it is, depends for its food on the remains of the yolk sac which is still attached to its abdomen when it is hatched, and it is only after this is absorbed that it begins to catch diatoms and copepods and find its nutriment in matters outside the fund of food provided for in the egg. All this time it is an inhabitant of the plankton.

Here then in the marine mammalia, the viviparous fishes, the marine reptiles, and in those fishes which lay large-yolked eggs we see what is known as the direct mode of development. Whether the period of incubation or gestation is long or short, that is relatively so, the little animal which is born out into the world is a fully-formed member of the species to which it belongs and it can be recognised as belonging to this species, for most of the characters of the adult are to be seen in the newly-born animal. This is because provision is made for the nutriment of the embryo and the process of development is carried to nearly its conclusion when birth takes place.

But this is not the case among the vast majority of the invertebrata of the sea. The direct mode of development is exhibited only very exceptionally by these animals. We find here that small eggs are produced, that these contain little food yolk, that development is a comparatively rapid process and that at the end of the incubation period a creature is hatched which is as a rule quite unlike its parents. It is a larva and unless the complete life-history of the species to which it belongs is known it cannot be recognised. On the land we have the same thing in the manner of reproduction of the insects. Here a "maggot" or larva is hatched out from the egg, and when the life-history is unknown, the mere form of this is no indication of the identity of the species which has produced it. Among marine invertebrata, as among the terrestrial insects, the mode of reproduction is an indirect one, that is the development of the adult conformation is not a simple continuous process but is interrupted by the formation of a larval stage, or of a series of such.

The characters of the plaice are well known to every one. At the beginning of the year this fish spawns about three hundred thousand eggs, which then drift about in the sea while undergoing development. At the end of a fortnight or so these eggs hatch and the little newly-born plaice are liberated into the sea. The parent fish is a flat creature coloured brown and red on one side but colourless on the other. It swims on one side of its body; its eyes are apparently both on the same side of its head; and its mouth is twisted to one side. But the newly-born plaice is round in cross section, it swims at first on its back and latterly with its back upwards in the proper manner, it is clear and glassy in appearance except that the tail is dotted over with black and some yellow pigment and it is quite symmetrical, one eye being situated on each side of the head. The plaice when it is recently hatched from the egg cannot be recognised as such, and we only know that it is a plaice because we know its life-history and we can easily watch the process of development. It is really a larval stage in the development of the fish, and the larva differs strikingly from the adult, because the egg from which it was hatched contains so little food yolk that the process of development has to be interrupted in order that the larva may grow and obtain food so that it may be able to complete unfolding of the adult structure.

Now it is such a process that is characteristic of the invertebrate population of the sea. Among the crustacea there is no animal which gives birth to one like itself. The eggs of these shellfish develop into larvae which are quite unlike their parents, at least when they are just born from the egg. Among the crustacea there are two principal larval forms, the *nauplius* and the *zoea*. The nauplius is the creature which is hatched from the eggs of all these shellfish with the exception of the Decapoda (the crabs, lobsters, prawns and their allies, where the larval form is the zoea). Even in these latter animals there is a disguised nauplius stage in the embryology of the zoea. When the egg of a crustacean hatches a nauplius issues and for a time this creature leads an independent life in the plankton. It is quite unlike the young of any of the higher animals, which is for a time cared for and fed by its parents, but it behaves from the time of its birth exactly like a fully developed animal except that it is not sexually mature. It swims about by itself and seeks for and finds its own food, behaving to the extent of its powers as a minute predatory animal.

The development of the crustacea is really more complicated than I have indicated, in that more than one larval stage is usually intercalated between the egg and the adult form. Thus the zoea larva hatched out from the egg of the decapod passes through various phases before the full adult characters are acquired by it. The crab zoea moults and passes into the *megalopa* phase. Both zoea and megalopa were originally described as adult animals and these names are those which were applied to them by their discoverers. So also in the nauplius of the copepods a series of disguised phases may be discerned in the development of this larva into the adult form. The nauplius of the cirripedes (the barnacles) lives in the sea for some time; then moults and grows more appendages, and finally a bivalve shell, and again drifts about in the sea for some time before it settles down on the objects of the sea bottom as the familiar organism which every one knows. Stomatopods, Ostracods, Isopods, Amphipods, Schizopods and Phyllopods all have their nauplius larva and a series of other more or less easily recognised larval stages appear before the full characters of the species to which they belong are acquired. In the peculiar parasitic crustacea which infest the bodies of fishes and other marine animals there is a much more complicated life-history, and a long series of larval phases may be included in the process of development by which the creature enters into its definite place in the metabolism of the sea.

Very few mollusca have a direct development. As a very general rule the egg which is spawned into the sea when the breeding season of the shellfish is reached is a minute organism possessing a larval shell and provided with a locomotory apparatus in the shape of a lobe of the body carrying a number of cilia. Cilia are the stiff contractile hairs by means of which the larva of nearly all marine animals progress. This *veliger* larva is as characteristic of the mollusca as the nauplius is of the crustacea. It swims about in the sea for a considerable period and then abandoning the planktonic mode of life for a benthic one it settles down on the sea bottom, attached to stones, &c., or in the mud, or in the sand, for the remainder of its life. The egg of the starfishes and sea-urchins develops into a larva which is as characteristic of these groups as the other larval forms we have mentioned are of their respective groups. The *pluteus* larva of the starfishes and sea-urchins is an inhabitant of the plankton and when it has lived long enough in this mode of life to become enabled to complete its development it begins to unfold the structure of the adult. The eggs of the worms hatch out as *trochospheres*, which are spherical or oval creatures possessing a very simple alimentary canal, two eyespots, a tuft of sensory hairs in front, and a circlet of cilia by means of which they swim. Successive joints or segments of this larval body bud off and the adult formation is gradually developed. Then if the parent has been a planktonic dweller the worm continues its life in that mode, but if it is a bottom-living animal it soon settles down to a life in the benthos.

Observe that there are larval forms which are characteristic of extensive groups of marine animals. The typical crustacean larva is the nauplius, that of the mollusc the veliger, the echinoderm larva is the pluteus, or some other characteristic form, and the larval form of the worms is the trochosphere. Although very considerable differences may be witnessed in the characters of the adults of the species belonging to these assemblages of animals it is nevertheless the case that the larval forms shew a very close resemblance. We explain this by assuming that in the evolution of the group there was an ancestral form which is now suggested by the larva. The memory, so to speak, of this ancestral form has been indelibly stamped on the species in that it is reproduced in the development of each individual. If an organism occurs in fossil form which resembles the larval form of any animal group we are justified in assuming that (if such a view is not contradicted by other evidence) this fossil form was related to the ancestors of the group in which our larva occurs. Thus *Limulus*, the king-crab, shews in its development a stage which is suggestive of the structure of the Trilobites, and it is generally held that these long extinct animals are in some way related to the king-crabs of the present day.

So far we have only considered the sexual mode of reproduction, but there are other methods by means of which marine animals multiply. Whole groups of animals reproduce by budding. Thus the polyzoa have in addition to the sexual mode, developed an asexual method of reproduction. These are animals which form colonies, and the sea-mat (*Flustra*) with its leaf-like fronds will be a familiar example of the group to any one who has looked at the débris cast up by the tide on the seashore. Starting with one individual produced by the sexual mode, a colony is formed by the formation of buds from this original *zoid*. So also with he zoophytes. These are plant-like organisms belonging to he coelenterata, and each colony is an organism which consists of a great number of separate units or polyps (or zoids). Here again we begin with one original individual which has budded to produce the colony. Corals are also colonial animals, and in the massive coral formations we see numbers of separate stony receptacles each of which is the house of a separate coral zoid. All of these are formed by the budding of one or more zoids. Siphonophores are also compound animals in which the zoids are formed by budding, and in these animals the division of labour among the zoids, which is also exhibited by some other compound animals, is well shewn. In a siphonophore there are assimilatory, sexual, locomotory and defensive zoids, all of them produced by the modification of originally similar individuals. In all compound animals the zoids are not entirely free from each other but all are connected together by a common flesh and are set upon a common skeleton. Each zoid may sometimes exercise the sexual function. Eggs and spermatozoa may thus be produced by each if they are hermaphrodite, or separate zoids may form these elements. Each colony begins in this way.

Then we have the method of reproduction by simple division which is generally characteristic of the protozoa, but is also exhibited by groups of animals much higher in the scale. In this mode the organism simply divides into two. But this process cannot continue indefinitely, and a phase in the life of each individual is attained in which reproduction by the union with another takes place. Here there is an alternation of generations Sexually produced and asexually produced series recur with more or less regularity. Alternation of generations is the rule in the vegetable kingdom and is characteristic of several animal groups Even in the higher sub-kingdoms of the animals there are evidences of an alternation of generations, though these are greatly masked. Thus even in the directly developed skate there are indications of the appearance of a larval stage in the embryology of the beast.

We may speak of the life-history of the majority of the animals in the sea as exhibiting various stages. There is (1) the embryonic

stage in which the animal is developing within the egg; (2) the
larval stage which begins when the embryo hatches out from the
egg and which may be reduplicated by the development of a
series of larvae; (3) the juvenescent stage in which the larva has
attained the form of the adult and can be recognised as belonging
to a definite species, even though its life-history has not been
followed out; and (4) the adolescent stage in which the young
animal has become sexually mature. This sequence is of course
only a general one and it is modified in all kinds of ways.

Death as the result of senile decay must be a very exceptional
event in the case of a marine animal. Usually the life of such is
determined by some catastrophe. In the sea the struggle for
existence is probably more severe than on the land. Every animal
has its own peculiar enemies, either predatory creatures which
prey upon it, or parasitic organisms which in their ultimate effect
are no less to be feared. Physical events may cause havoc in the
sea. Violent alterations of temperature may lead to the death of
hosts of creatures, while great changes in the salinity of the sea-
water may be no less fatal.

Countless millions of pteropods must be destroyed by the
whales of the northern seas; porpoises destroy hosts of herring,
cod, whiting and other fishes; roving sharks and dogfishes, either
singly or in shoals, must at times produce devastation among the
bottom-living fishes of sea areas; cod which are themselves the
prey of porpoises devour great numbers of fish such as herrings,
and crustacea such as hermit crabs, &c.; plaice and flounders eat
enormous numbers of cockles, mussels and other small shellfish,
and densely populated beds of these molluscs are at times
decimated by hordes of starfishes; pelagic fishes like herrings and
mackerel feed to a great extent on swarms of copepods and other
planktonic crustacea, and 20 millions of *Ceratium* have been
estimated in the stomach of a single sardine. And so on through
the whole marine animal kingdom. Sea birds prey to an astonishing extent upon molluscs and fishes. Finally man, the destroyer,
contributes his share to this incredible massacre, for whatever is
useful in the sea is caught by him to the full extent of his powers,
aided by fishing machinery; and not only useful animals but also
those which are in association with these are destroyed. But it is

only in the case of the larger fishes and the rarer marine mammalia that man's influence as a destructive agent is seriously felt. As a destructive agent he has far less effect on the abundance of most useful marine animals than is at first apparent.

But no less astonishing is the incredible fecundity of marine organisms and the variety of contrivances for the protection of both immature and adult animals which the struggle for existence has evoked. If the amount of destruction is great no less great are the recuperative powers of a marine species, and indeed it is only by measuring the latter that one can arrive at an estimate of the amount of the destruction. A large turbot may spawn nine millions, a cod five, a flounder one, a plaice three hundred thousands of eggs during each breeding season. Other fishes are almost as fertile. Even the relatively slowly-breeding crustacea produce great numbers of eggs: crabs, lobsters, prawns and their allies produce several thousands of eggs at each spawning. Most molluscs are very prolific, and though no one has attempted to estimate the number of eggs annually produced by a mussel, yet a glance at the enormous number of eggs that can be taken from the ovary of this animal on the point of a knife will convince one that it must be very great. Starfishes and their allies are very fecund. Bottom-living worms often deposit their eggs in cocoons and each of these may contain an astonishing number of ova. Speaking quite generally benthic animals are more prolific than pelagic forms, but even in the planktonic worms hundreds of eggs may be carried by one individual. Pelagic crustacea also are less fertile than bottom-living forms, but they have advantages in the more abundant opportunities for the distribution of their eggs and larvae. The powers of reproduction of the compound polyzoa and zoophytes are astonishing. In these animals large numbers of zoids are produced by the budding of an originally single individual, and of these zoids, which together form the association known as the polyzoon or zoophyte, a large proportion at least are reproductive and produce eggs which are liberated in the sea to develop into free-swimming larvae, which either settle down on the sea bottom to reproduce another colony or may themselves bud asexually or again produce eggs. Hardly a marine animal but harbours some parasite and animal parasites above all other organisms are incredibly fertile. In a single tapeworm inhabiting the intestine of the skate or cod there may be hundreds of separate segments or joints and each of these is a sexual unit and may produce many hundreds or even thousands of eggs. When we consider the unicellular organisms we are no less impressed with their fertility. A very small mass of diatoms put into a dish with a little sea-water and a piece of weed and exposed to the sunlight for a day or two multiplies to an astonishing extent. All unicellular organisms which reproduce by simple division do so at times with great rapidity. Given the proper conditions of nutrition and a few infusoria contained in a little mud will in a very short time have produced an incredible progeny. One single marine bacterium will even at the ordinary temperature multiply to the extent of several millions in the course of a week or two.

On the other hand the life-histories of some marine animals are complete antitheses to those indicated above. Whales, seals, porpoises and other marine mammalia are like their terrestrial allies slowly-breeding creatures. The viviparous dogfishes produce few offspring in the course of the year. Those fishes, like the skates and rays, which lay large yolked eggs only deposit a few dozen at the most in the breeding season. Pelagic worms and crustacea are not nearly so prolific as the demersal forms. The viviparous invertebrata produce few embryos in the year.

Now it is not difficult to explain why it is that such differences in the fertility of marine animals should exist. It is an apparent paradox that while a turbot produces annually nine millions of eggs, and a ray only a dozen or two, the ray should be much more abundant and more widely distributed than the turbot. But there is no necessary relation between the fertility of a marine organism and its abundance. "No fallacy," says Darwin, "is more common among naturalists, than that the numbers of an individual species depend on its powers of propagation[1]." All that the fertility of such fishes as the turbot and cod, or such molluscs as the mussel, or crustacea like the shore-crab, indicates is that the destruction of these creatures at some stage of their life-history is enormous—just as enormous as their fertility is great. And conversely, the fact that a fish like the ray or dogfish, or a

[1] *Voyage of the Beagle*, Chap. IX.

crustacean like the lobster, produces comparatively few offspring indicates to us that some provision is made for the rearing, and protection from enemies, of the eggs and larvae. Natural selection has taken two ways of combating the destruction of animals in the sea. On the one hand enormous numbers of eggs are produced only to be destroyed, while a few survive to maintain the species; and on the other hand few eggs and offspring are brought into the world and all kinds of devices are elaborated for the protection of these.

Thus we find that fishes like the cod, whiting, turbot and herring shed their eggs into the sea and take no further thought for them. The ova of the first three fishes are shed into the sea and drift about in the plankton, and though they are so pellucid as almost to be imperceptible to the human eye, yet it is probable that many pelagic animals feed upon them. Herrings deposit their eggs on the sea bottom and then abandon them, and soon after hordes of whiting come and gorge themselves on this nutritive material. Herring, whiting, cod and turbot while they are in their larval stages are among the most helpless of creatures that swim in the plankton and while they are still microscopic in size many pelagic animals must feed upon them. Then, with hosts of other young fishes, they are stranded on the shore by the receding tide, and being left in shallow sand pools, which soon dry up beneath the heat of the sun, they perish miserably. Even when these fishes have attained the age of juvenescence they are devoured by other animals. Dozens of whiting furnish only one meal for a porpoise. Cod prey upon turbot, whiting and herring; and the whiting is not infrequently cannibalistic in its proclivities. Skates, rays and dogfishes are catholic in their tastes and prey upon all four of the fishes instanced as examples of the reciprocal destruction of fishes by each other. But the skates, rays and dogs are themselves fairly free from enemies, and only the porpoises prey on the dog-fishes and man on the skates and rays. The eggs of the two latter fishes are enclosed in hard, horny indigestible capsules and are not in danger of being eaten by other animals. They are laid on the sea bottom and are not in danger of being stranded on the shore. So too with the eggs of the oviparous dogfishes. All these eggs are large and contain much food yolk, and when the young fish hatches out from the egg it is already a large and predatory creature which is able to seek shelter and to find food. The viviparous dogfishes have even greater advantages. The young, because of their mode of development, escape the dangers of embryonic and larval life and when they are born are active creatures which are able to avoid some at least of their enemies. The intelligent and pugnacious lobster carries its eggs attached to the abdominal feet, and during the nine or ten months required for the incubation of these they are immune from enemies. Even the fisherman is prohibited by law from taking the nursing lobster or crab. Generally among the crustacea the eggs are carried by the mother during the incubatory period, and there are hosts of devices by which these, and other marine animals, are enabled to carry and protect their eggs and larvae: such are brood-pouches and chambers, egg-sacs, &c. Some of the fishes devote what we must call purposeful and intelligent care to the fostering of the eggs and young. The male pipe-fish carries the eggs in a brood-pouch; the lumpsucker guards the mass of ova laid by the female, and keeps it aerated by blowing water upon it from his mouth, or by waving his tail backwards and forwards in front of the egg heap. Often his devotion is the cause of his death, for the eggs may be deposited in a rockpool which may be exposed by a spring tide and the unhappy parent may then be attacked by predatory birds or small boys. The fifteen-spined stickleback builds a nest out of sea-weeds and cements these together by a viscid fluid secreted by him at the spawning time. In this nest are laid the eggs of one or more females. And so all through the marine fauna. Fully to relate the methods developed by natural selection for the protection of the eggs and young stages of fishes and other marine animals would require a large volume.

Now we may return to the plankton. We have seen that this contains both permanent and transitory components. Hosts of benthic animals and plants while they are still in the larval, or sporing phases, live among the drifting microscopic life of the sea and then disappear, having assumed some one of the other modes of life. If then we consider only the transitory life of the plankton we should expect to find that both the nature and amount of the organisms present should vary with the season, and observation indeed shews that this is the case. Even the permanent plankton varies with the season. In the sea, as on the land, there is seed-time and harvest, and with the spring a multitude of life comes into existence to pass away in the autumn and winter. I leave aside in the meantime a consideration of the causes of the variation in abundance and nature of the plankton. We shall see later that enquiry into this carries us away from the traditional province of the biologist into that of the physicist and astronomer. Very probably many of the larger variations in the plankton are caused by cosmical events.

It is possible to construct a "calendar of the plankton[1]." If we make periodical hauls with a tow-net in any small sea area once a week or oftener throughout the year it is easy to see that the fauna and flora of the plankton do not vary fortuitously, but there is a very definite order of succession in the nature and abundance of the organisms found throughout the year. Fishes, molluscs, crustacea, echinoderms and other creatures which live in the sea all have their spawning seasons. The nature of the seasons may delay or hasten these spawning times; exceptional gales or unusual periods of fine weather may have great influence and produce changes in the abundance of the permanent plankton, confusing the order of appearance of the transitorily occurring organisms, but when the observations of a number of years have been accumulated one knows roughly when to expect the appearance of the usually occurring things. I will consider here the main changes in the order of the plankton which may be observed in the sea off the west coast of England.

There is never absolute sterility in the sea. Wherever and whenever one fishes with the tow-net life is to be found. But at the beginning of the year the plankton is not abundant, that is with relation to the other seasons. The permanent constituents are generally scarce and the great outburst of life, which is due to the spawning of the fishes and other bottom-living or nektic creatures, has not yet taken place. But very soon, as the intensity of sunlight increases and the amount of food matter in solution in the sea begins to accumulate, the diatoms multiply at an increased rate.

[1] McIntosh, *Ann. Rep. Scottish Fishery Board* for 1889. See also Garstang, *Journ. Mar. Biol. Ass.* Vol. III., 1893–5.

Previously many of these plants have formed resting spores, and are lying dormant. Genera like *Biddulphia*, *Coscinodiscus* and *Chaetoceros* appear about the end of February, at times in great profusion, so that the tow-nets may occasionally contain little else than these organisms. The fishes begin to spawn, and the copepods too increase in numbers. But the changes which characterise the months of February and March are (1) the increase in the diatoms, and (2) the appearance of fish eggs and larvae. The outburst in diatoms is due to meteorological changes principally, and it effects a part of the plankton which is already present in the sea as such, but the fish eggs and larvae result from the multiplication of animals which are nektic in their habits. The diagram below has been constructed from observations of the appearance of fish eggs in Cardigan Bay for the years 1905-6[1]. It shews that even at the end of the year plaice begin to spawn. Then follow the rockling

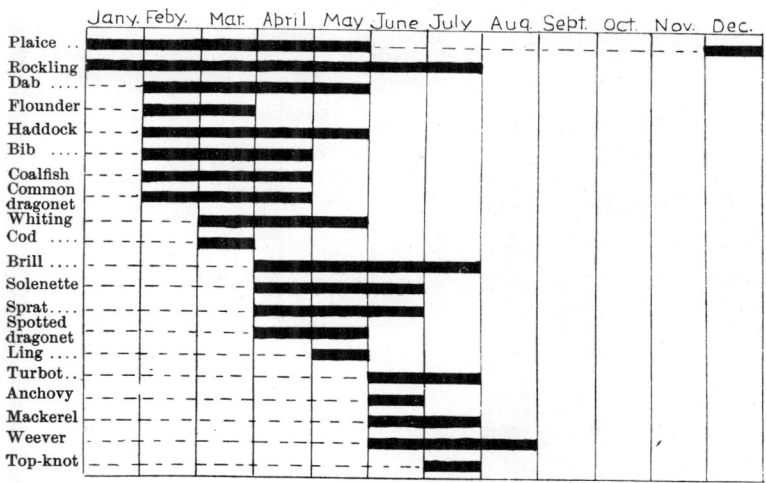

Fig. 22. The succession of fish ova in the plankton of Cardigan Bay, 1905 and 1906.

(*Motella*, probably more than one species), flat-fishes like the flounder and dab, gadoid fishes like the haddock, bib and coalfish, and the common dragonet. Cod and whiting begin to spawn in March and the brill, solenette, sprat and spotted dragonet complete

[1] Scott, *Annual Reports Lancashire Sea-Fisheries Laboratory*, 1905 and 1906.

the list of fishes the eggs of which have been taken in the tow-nets here in the spring months.

Simultaneously, the larvae of many benthic and planktonic creatures appear in the tow-nets. The diagram below has been constructed from observations of the same series of collections as that from which the distribution of fish eggs has been made out,

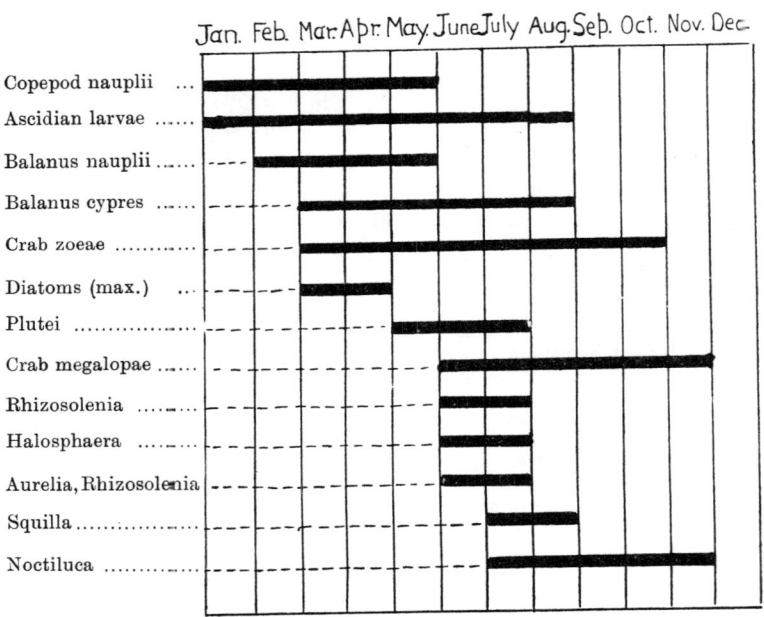

Fig. 23. The rough succession of some planktonic organisms in the sea off W. England.

and it shews the appearance in the course of the year of the larvae of some of the commoner animals of the plankton. Copepods are now and have for some time previously been carrying egg-sacs, and the eggs in these begin to hatch out so that the nauplii of these crustacea begin in January to appear in the plankton. But they are not abundant, for the copepods produce comparatively few eggs and, as a group, spawn over a prolonged period of time. We find their nauplii in the sea until the end of May. Ascidian eggs containing developing embryos are also observed at the beginning of the year and last until the autumn. A more striking change is the advent of myriads of nauplii of barnacles (*Balanus*) which at times are found in the tow-nets to the practical exclusion of almost everything else. Great numbers of these animals must spawn simultaneously and very rapidly, for they appear to produce swarms of larvae. After a short free-swimming existence these nauplii moult and we then find that the tow-nets contain the bivalved cypris stage of *Balanus*. These latter larvae linger on until the end of the summer, but in April and May most of them must have begun to adopt their adult habitat, for we find then that swarms of young *Balani* have settled down on the stones, shells, piers, sea-walls, &c.; and on the hulls of the boats of the fishermen, in spite of anti-fouling paint, assiduous scrubbing and doubtless much profanity.

At the beginning of March the crabs begin to spawn. The distinction of the different crab zoeas is a matter of great difficulty; certainly there is much individual variation in the time of the spawning; and the larvae have, as such, a long life. Therefore we find that most tow-nettings taken until the end of October contain zoeas. At times the eggs carried by numbers of crabs must hatch almost simultaneously, for we find tow-nettings which contain little else than these larvae.

May begins a new season. The diatoms have hitherto been prominent plankton organisms, often to the practical exclusion of everything else, but from now for a few months they begin to decrease in numbers. The summer spawning fishes begin to shed their eggs. Ling, turbot, anchovy, mackerel, stingfish (*Trachinus*) and topknot (*Zeugopterus*) spawn in May, June and July. The sole too spawns in June, though the eggs of this fish do not appear in the inshore waters of Cardigan Bay and are not represented in the diagram. Plutei of echinoderms appear in May for the first time. Sometimes the beautiful long-spined plutei of the heart-urchin (*Echinocardium cordatum*), a species which lives in the mud in inshore waters, is so abundant as to crowd out other organisms from the tow-net. In June the crab zoeas are undergoing their metamorphoses into the Megalopa stage, and swarms of these larvae appear at the surface of the sea, and also at the bottom, for dabs may be taken with their stomachs distended with these creatures. Megalopas are long-lived and November sets in before they disappear from the plankton. The beautiful Erictheus larva of the Stomatopod *Squilla* appears in July and August. In July the sea often becomes brilliantly phosphorescent and the tow-net shews that this phenomenon is due, as a rule, to the protozoon *Noctiluca*, an organism which is the usual cause of phosphorescence in the sea off the west coast of England.

The summer months are also characterised by the appearance of two series of organisms which are an intolerable nuisance to the fishermen and naturalists. Peculiar algae (*Tetraspora* and *Halosphaera*) which secrete mucoid capsules, appear in great abundance, clogging up the pores of the tow-nets, and drying on the trawl nets, producing an unpleasant odour. Great swarms of medusae float on the surface of the sea. At first *Aurelia* is the common one, but later on the large "cabbage-bleb" (*Rhizostoma*) makes its appearance. One may sail for miles through a swarm of *Aurelia* so densely packed together that the sea has a uniform reddish colour. When the trawl is hauled it is full of this jelly-fish and great labour is often required to clear it, and when at length it is brought on deck much shovelling is necessary before the slobbery mass of Aurelia is thrown overboard. Large *Rhizostomae* are often seen drifting about in the water, each with a little crowd of gadoid fishes of small size sheltering underneath the bells, doubtless seeking the shade and also protection from enemies, to whom the stinging jelly-fish is a thing to be avoided.

During the summer months the diatoms have not been so abundant as formerly, though they are always present to some extent. *Rhizosolenia* is often abundant, being perhaps more characteristic of the summer months than the other diatoms mentioned. With the autumn and winter months the diatoms begin to flourish again. A second period of increased multiplication occurs and from then to November they again become abundant. We shall see that this second maximum, like the first one, depends on the accumulation of dissolved food matter in the sea, perhaps ultimately on the reduction of the temperature.

Towards the end of the year the plankton as a whole falls off both in quantity and variety. The larval forms have mostly become metamorphosed and have adopted bottom-living habits or have taken to a free-swimming life in the sea as active animals.

The perennial plankton continues. But the summer visitors among the copepods and other crustacea have departed; also the large medusae and other coelenterates, the pelagic algae, and many other permanent planktonic animals and plants have gone. The ubiquitous copepods such as *Paracalanus, Pseudocalanus, Acartia, Centropages, Temora, Oithona,* and others remain in restricted quantity. The ctenophore *Pleurobrachia,* the pelagic worm *Sagitta,* and the ascidian *Oikopleura* remain also. *Ceratium* is nearly always present. But all are less abundant during the last month of the year. Finally, at the beginning of the year the diatoms again begin to increase attaining their maximum in the spring.

There is thus a regular succession of organisms in the plankton of a sea area. In that just described we see that the main features of this succession are:—(1) relative scarcity at the beginning of the year; (2) an outburst of diatom life in the late winter and early spring; (3) the appearance of countless myriads of fish and invertebrate eggs and larvae in the spring and early summer; (4) a decrease in the abundance of the diatoms, and the gradual disappearance, from the plankton, of the eggs and larvae; (5) the appearance of swarms of medusae and other coelenterates in the summer; (6) the reappearance of diatoms in the late summer and autumn; and (7) the scarcity of the plankton as the winter begins. Throughout all there are a number of forms of life which remain as permanent inhabitants of the area.

Just in the same manner the fishes and the bottom-living invertebrata have their phases of abundance and scarcity. In the Irish Sea cod appear in the winter and spring and spawn in the sea offshore. Hordes of whiting visit the offshore fishing-grounds in the spring. Flounders and small plaice migrate from somewhere to the shallow inshore waters in the spring. Soles appear in the deeper waters off the land in the summer. In the summer and autumn plaice visit certain fishing grounds for a time and hake migrate into the deepest parts of the sea from the south. Fishermen are well acquainted with these periodic appearances of fishes and regulate their operations accordingly; and just as in the case of the plankton there are ubiquitous fishes, such as the dab, which are to be found everywhere and at any time of the year. No doubt corresponding series of changes occur in the case of the bottom-living invertebrata, but these have not been studied to the same extent as have the plankton and the fishes.

All throughout the year there is a more or less regular change in meteorological conditions. The temperature of the air is continually changing: violently at times, because of changes in winds, as cyclone succeeds cyclone, or are interspersed with the more welcome anticyclonic systems; but throughout the year rising generally to its maximum and again falling to its minimum. The temperature of the sea too undergoes much the same variations, but these are more regular than in the case of the atmosphere. Sometimes the difference between sea and air temperatures is relatively great, sometimes relatively small. The duration and intensity of sunlight vary from day to day; and to some extent at least there are variations in the salinity of the water. All these changes in physical conditions react on the life processes of marine organisms, but the primary cause is the annual change in the temperature of the water. Just as there is a great outburst of vegetation on the land in the spring and summer, so there is a spawning season in the spring and early summer in the sea, and in the case of most marine organisms this habit of reproduction during these months has been stamped upon them by heredity. We thus find that the hinge round which most seasonal variations in the abundance of animals in the sea turn is the annual reproductive phase sometime or other during the first six months of the year.

There is only food for a limited (though of course very great) mass of life in the sea, and any change which is favourable to the multiplication of the individuals of one species must necessarily react on other organisms. The sudden production of the vast numbers of larvae, which result from the spawning of hosts of fishes and invertebrates in the spring, must lessen the numbers of diatoms, protozoa, or micro-crustacea on which those larvae feed. On the other hand, the appearance of the eggs and larvae means that food is provided for other predatory animals: thus haddock follow the spawning herring and gorge themselves with the eggs which the latter fishes deposit on their spawning grounds, and to such an extent does this destruction take place that hundreds of boxes of haddock have been landed by fishermen, and all these fishes have had their stomachs full of herring eggs. If by any circumstance one species receives an advantage then some other necessarily suffers. Thus it is maintained by fishermen that the operation of the Wild Birds Protection Acts has been to cause a diminution in the abundance of cockles in some localities, since increased numbers of gulls were spared to eat these molluscs. Fluctuations in the abundance of planktonic organisms in the sea, which are smaller than the great annual periodic changes which we have been considering, are often to be traced to such minor disturbances in the "balance of life." In the sea there is always a certain relation between the numbers in groups of species, and though at any one time there is an equilibrium, still this equilibrium is a continually changing one; and changes in the physical conditions of the sea—which changes are fundamental ones—must disturb the balance.

CHAPTER V.

THE SEA-FISHERIES.

INTO this complex and delicately adjusted struggle for existence man now throws the weight of his influence, and as the captor of hosts of marine organisms becomes a disturbing factor of importance. Among the organisms which populate the seas of the globe are very many which are useful to him, either as food, as the materials for adornment, or as the raw stuff for use in the arts and manufactures. Whole groups of fishes such as the cod, plaice, ling, haddock, herring, dab, flounder, witch, skate and ray, dogfish, and others are food for the humbler members of the population, while others such as the sole, turbot, salmon, halibut and brill provide luxuries for those who are able to afford them. Among the crustacea the crabs and shrimps are the food of the more penurious, while the lobster adapts itself to the more affluent. Periwinkles, mussels, and cockles are luxuries suited for the narrower purses. Oysters and turtles conjure up images of aldermanic junketings. Other more obscure organisms find their places in the dietaries of many of the peoples of the world. Edible sea-urchins (*Echinus esculentus*) and holothurians (Beche-de-Mer and Trepang) are eaten in some parts of the world. Polychaete worms too form an article of food to some primitive peoples. From very early times the humble, though succulent "dulce" (*Iridaea edulis*) has been eaten in this country, and other marine algae are also articles of food. Everywhere the sea is the repository of stores of food for man. Not only so but other important economic products are also yielded by the blue water. Numbers of shellfish which are not taken for food are of considerable value. The Mytilidae of warmer seas furnish the "Orient Pearl" of the jeweller, and the shells of the larger *Margaritiferae* are so valuable for the mother-of-pearl that they contain that they are worth some £200 a ton. The fresh water mussels of our rivers (*Anodonta* and *Unio*) yield pearls which have occasionally some value as gems; and the common edible marine mussel too furnishes pearls which had at one time a use in commerce. Some of the larger tropical mollusca are provided with a byssus which is so large that its fibres are spun into cloth, which is used in those parts of the earth where clothing is worn for its primary purpose alone. The imperial purple of the Romans was yielded by the molluscs *Murex*, *Purpura* and others; and the sepia of the artist is obtained from certain cuttle-fishes. Many molluscs, both bivalves and univalves, are used as ornaments, and from time immemorial the cowries (*Cyprea*) have served as a primitive form of money. Coelenterates furnish few products of use to man, but the precious red coral of commerce is a notable exception. The whales and seals are, of course, very valuable, the former for the oil and whalebone, and the latter for their skins. Many useful products are yielded by parts of fishes caught primarily for food, thus the swim-bladders of fishes are used to obtain isinglass, and other parts of the piscine anatomy are employed for the manufacture of the coarser fish-glue. If no other use can be made of fish offal it is converted into manure, and large quantities of starfishes find this application in commerce, a use to which the rejectamenta of our fish markets is also destined. Seaweeds are also made into manure, and not so long ago all the bromine and iodine of commerce were made from the kelp obtained by burning the marine algae.

Man has exploited the population of the sea from the earliest times. Long before the origin of the arts and manufactures he fished and hunted, and as the human animal increased in numbers he became the most formidable enemy of those other animals with which he at first competed on more or less equal terms. Man, the hunter and sportsman, soon began to decimate the land, so that in all civilised countries the larger and more slowly breeding animals have become scarcer and scarcer, or have disappeared, as has been the case with the elephants, bisons, deer and wolves. Even the smaller and less valuable mammals like the hares, otters, badgers, foxes and others are now rarities in densely populated countries. The salmon would long ago have practically disappeared from the rivers of Britain if the value of this fish to the sportsman had not been prolific of legislation for its protection. Eagles, ravens and other birds have yielded to man's destructive instincts, and in other countries the larger flightless birds such as the ostrich, cassowary, emu and kiwi, are slowly disappearing or require to be cultivated or protected by legislation. The moa, great auk and dodo have become extinct, but these were probably species which were passing away from natural causes before man came on the scene. In our own country many wild birds are the objects of protective legislation, mainly in the interests of sport. Numbers of birds and mammals which are useful or indispensable to man are now bred and domesticated. Thus man's influence has been very powerful in the case of the larger terrestrial creatures, and to some extent this is the case in regard to some marine species. Steller's sea-cow (*Rhytina*) has become extinct since 1786[1] and the manatee and dugong are becoming rare. It is likely too that the whales and seals will in the future become very scarce if protection is not invoked in their behalf. But in the sea the area to be exploited is so vastly greater than on the land that the influence of man is only just appreciable. The fact that he is still the hunter of marine animals shews that this is still the case. Fishes are not bred or domesticated because these measures are necessary in order to preserve them in sufficient numbers to afford food or ornament. With few exceptions the sea is exploited in the twentieth century essentially as it was in post-tertiary geological times. If we now cultivate oysters and a few other truly marine animals it is for convenience, or because greater commercial gain is so obtained, and not because it is necessary to do so in order that these animals may be preserved in sufficient numbers to form the material of a fishery. The "Harvest of the sea" is a metaphor that appeals only to the uninstructed. In the ocean nature sows the seed and man reaps where he has not sown.

But nevertheless the fishing industry has undergone very much the same degree of elaboration and specialisation that some

---

[1] *Rhytina* was discovered on Behring and Copper Islands, in the North Pacific, by Behring and Steller in 1741. Forty years afterwards the last individuals of the species were extirpated by the hunters and traders who followed on the track of the explorers.

manufactures have experienced in the strenuous struggle for wealth of the last century. Though it is essentially still the art of the hunter yet the methods evolved for the capture of marine animals have been greatly improved during the last few decades of that era. The fishermen have directed a great amount of observation towards ascertaining the habits of fishes and other economic animals with a view to their capture in increased numbers and fishing apparatus is continually being improved.

The methods of modern commercial fisheries are of course based on a study of the habits of marine creatures. Bottom-living nektic animals are caught by means of apparatus which work on or near the sea bottom: these are trawls, dredges, lines, bottom set nets and seines, crab and lobster pots and other devices. Truly benthic animals are caught by the same apparatus, or by other more special forms of fishing gear such as the tangles or swabs which are used to obtain the red coral of commerce. Pelagic animals are caught by drift nets, seine nets, and lines. Planktonic organisms have as yet no use in commerce and the fisherman is as a rule ignorant of their existence, except such as form a nuisance to him.

It is quite impossible to do more than to indicate the principal methods of fishing pursued in the seas round the British shores. An adequate account of fishing gear has yet to be written, but at any rate we are concerned here not so much with the study of the fishing instruments from the point of view of the professional fishermen as with their interest to the naturalist in a study of the general conditions of life in the sea. So I will notice only the principal types of fishing gear. These are modifications of the dredge, seine net, trap and line. The trawl and line for the bottom-living nekton and benthos, and the seine and line for the pelagic nekton are the types of fishing apparatus. The trawl is now incomparably the most important of the instruments used in sea fishing, and the history of the development of the fishing industry during the last quarter of the nineteenth century is that of the continued improvement of trawling vessels and their fishing gear. All kinds of trawling vessels now ply the seas. There is the modern steam trawler, a relatively large and powerful vessel commanded by an expert navigator (for she has to traverse an extensive sea-area) and equipped with electric light and all the refinements of modern naval architecture which have strictly utilitarian objects. At first the steam trawler was an obsolete tug-boat but now she is a steam vessel *sui generis*. She is often fitted with automatic sounding apparatus, for the trawl is now used in much deeper water than was the case not so many years past. Even a short time ago 50 fathoms was regarded as the limit at which it was possible to use this fishing instrument, but now the trawl is often used in water of 100 fathoms, and there is little doubt that this depth will soon be greatly exceeded. Simultaneously with the improvement of the vessel the trawl itself has been perfected. The modern steam trawler carries two trawls each of about 100 feet spread. While one of the nets is on the sea bottom the other is being made ready for shooting, and the fish just taken are being gutted and stowed away in the fish hold, so that the fishing is carried on continuously. The trawl is hauled by means of two strong steel wire ropes which are wound on the drums of the steam winch, and a modern steam trawler may carry about a mile of wire on each side of the windlass.

The traditional trawling vessel is the smack, a yawl-rigged vessel of about 80 tons burden, and manned by four "hands." She almost always carries a small steam-engine for the purpose of hauling the net, and working the sails, but the latter are the sole means of propulsion. Not having the length of side nor the room of the steam vessel the smack carries the older beam trawl, but this is nevertheless a large net which has a spread of about 50 feet of sea bottom. There is little doubt that the smacks are passing away. Every year sees more steam trawlers on the fishing register, and as the smacks are lost, or become obsolete, they are seldom replaced by vessels of their class. When they disappear the sea will have become less picturesque, and the population of our seaside towns will, to some extent at least, have deteriorated.

Then we have the second class fishing boats. These are half-decked vessels of about 10 tons in register and about 30 to 40 feet in length as an average. Their crews consist of two men, a man and a boy, or one man alone. They carry a trawl which does not usually exceed 30 feet in length. Sometimes they may carry two shank nets, or even four of these fishing instruments: in the latter case the vessel will tow one shank net from over each quarter, and one from a boom rigged out over each side. The half-decked boats, nobbies, or by whatever name they are called, are very often very pretty and substantial little vessels capable of standing quite hard weather. Great care is often exercised over their design and construction, for they are used not only for fishing, but also for racing and pleasure sailing. On many parts of the British coasts an open boat is used for trawling. Sometimes the trawl is worked from a cart, and the motive force is in this case derived from a horse which wades in the water up to its middle. The cart usually drags two shank nets. One sees this amphibious apparatus about the Southport shore, but in few other localities.

All kinds of fish are caught by the trawl net. Even herring, which are pelagic fishes, are often taken in numbers by the trawl, though this is an instrument which fishes on the sea bottom as a normal procedure. In shooting the otter trawl of the steam vessels the net is dragged through the water for a short time while the ship steams slowly ahead. If then the net is shot among a shoal of herrings or other pelagic fish the latter are meshed and caught. Lately this method of working the otter trawl has been used by the Fleetwood trawlers who happened to be on their usual fishing grounds while herring were in the neighbourhood. But with this exception the fish caught by the trawlers are such as live at, or near to, the sea bottom. The steam trawlers on the east coast of Britain catch vast quantities of haddock and indeed this fish appears to be their mainstay. Within the North Sea the principal fishes caught by the steam trawlers are cod, haddock and plaice, and then follow whiting, skates and rays and a host of other fishes in less abundance such as dabs, lemon soles, witches, gurnards, coalfish and many others. Outside the limits of the North Sea, cod form the principal fish caught by the steam trawlers of the east coast ports and then follow haddock and plaice. Steam trawlers from the west coast fishing ports (Fleetwood, Milford and Liverpool) do not fish in the North Sea but in the sea to the south and west of Ireland, on the west coast of Scotland, and at times in Icelandic waters. Deep water is frequented, and we find that the hake, which is a fish living on the deeper fishing grounds, is the most abundant fish in the catches made by these vessels. The smacks are of necessity restricted to the home waters and we do not find them fishing outside the North and Irish Seas and the Channel. Plaice, whiting, dabs, skate and ray form the bulk of the fish caught by them. At one time prime fish, that is soles, turbot and brill, were the fish sought for both by steamers and smacks, and most other kinds of fish were regarded as "offal." Although it is still considered a very desirable thing to catch large quantities of prime fish they have not the relative importance in the hauls of the fishing boats that they used to have. The demand for fish food has greatly increased since the days when haddock were regarded as offal, and the fishing vessels are now obliged to bring in to the markets fishes which formerly they rejected, such as angler fishes (*Lophius*), dogfishes and catfishes. Prime fish are mostly caught in the North Sea, the Irish Sea and the Channel (though notable quantities have been taken on the newer fishing grounds) and they form a relatively more important proportion of the catches of the smacks than of the steam vessels.

The half-decked boats are more concerned with catching shrimps and prawns than the fishes, though considerable quantities of the latter are often caught by them. These little boats are very versatile in their choice of employment, and shrimping, prawning, fish-trawling, lining, mackerel fishing, drifting, racing and pleasure sailing all are practised by them in their season. Nothing comes amiss to them. The shrimp trawl and the fish trawl are interchangeable according to the circumstances of the moment. Prawning (for *Pandalus*, the "pink shrimp") is rather a specialised form of fishing and is carried on more or less constantly by a number of half-decked boats in some localities, near Fleetwood for instance. The catches of the prawn and shrimp boats are often of much interest to the naturalist, for the restricted size of the mesh, and the employment of the net in what is perhaps the most densely populated part of the sea, give a catch which usually contains many organisms of interest. To the naturalist the hauling of one of these nets is always an operation of an instructive kind. Hosts of the smaller immature fishes such as dabs, plaice, soles, whiting, codling, and other gadoid and flat fishes are caught in numbers which depend on the locality, the season and the weather. Small inedible fishes such as sting fish, dragonets, solenettes, sand-eels, pogges, suckers, sticklebacks, gobies and many others are taken as a rule. The invertebrates of the shrimp and prawn trawls are always very abundant, and are often varied in their nature. Crabs and starfishes are nearly always abundant when the shrimp trawl is used, and these creatures are of course represented also in the catches of the prawn boats. Since the latter work on rougher ground than the shrimp boats, weeds and zoophytes are also taken, with the usual invertebrate fauna which is associated with these organisms. Greater novelty is thus exhibited by the prawn, than by the shrimp trawls.

Next in importance to trawling comes lining. The employment of lines is probably an older form of fishing than trawling or seineing, and even now when the liners are very gradually being superseded by the trawlers the former are still of considerable importance so far as the supply of fish is concerned. But the older line boats were vessels of very unpretentious build and equipment, whereas the newer line boats are vessels of much the same kind as the modern steam trawlers. The long-line used by the newer lining steamers is a fishing instrument of some magnitude. Each line carries a number of short pieces of rope to which the hooks are attached. These are the "snoods," and they are fastened to the line at intervals of about six feet. The baits employed vary according to the nature of the fishing and the convenience with which the bait animals are obtained. Shelled mussels, herrings, whelks, squids and other animals are the more common baits. They are put on the hooks ashore and the baited lines are then coiled up neatly in tubs, the separate layers being kept apart by bent-grass or some other substance. As line after line is shot overboard the ends are fastened together until as much as six or seven thousand yards of rope are on the sea bottom. The whole is weighted and buoyed at intervals. Usually it is shot one day and fished the next one. Often the bait is taken by some small fish and then the latter is taken when on the hook by some larger piscivorous fish. Often the line cannot be fished for several days on account of bad weather, and then hosts of squids or dogfishes may attack the fishes on the hooks and spoil them. Quite a different assortment of fish are taken by the long-lines. In the North Sea cod, ling, haddock, skates and rays are, in the order named, the principal fishes caught. During the last few years steam liners have increased greatly in number and now go beyond the limits of the North Sea proper to the north and north-east of Scotland, where the halibut is the most valuable fish caught. Then come cod, ling, skates and rays, tusk (*Brosmius*), and other fishes in less abundance. I am speaking so far of the larger line boats which are comparable in their equipment and personnel with the steam trawlers and smacks. But all round our coasts short lines are set, and this form of fishing carried out from small boats is often of considerable local importance.

Quite different methods are followed in the pelagic fisheries. Here we have to catch nektic animals which inhabit all strata of the sea from the bottom to the surface. The principal pelagic fish is the herring, and then come the mackerel, pilchard and sprat. Herring fishing lasts all the year round, for the great shoals visit in a rough sort of succession all parts of the British coasts from Stornoway round the north of Scotland and down the east coasts of Britain. Smaller fisheries take place at various parts of the coast; in the Firth of Clyde, round the Isle of Man, in the English Channel and to a limited extent off the coast of Wales. In nearly all localities herrings are caught by drift nets, which are perpendicular walls of netting, connected together to form "fleets" or "trains" each of these being about 8 yards in depth and often two miles in length. The nets are weighted and buoyed and sunk to the depth at which the fisherman judges the shoals are situated. They are not attached but drift with the tide, and the fishes striking against them are enmeshed by the gills. Sometimes (in the Clyde) a seine net, "circle net," or "trawl net" is shot round the shoal, or part of it, and is then hauled enclosing the fish[1]. Mackerel are caught either by nets or they are trolled for by light lines sunk a little below the surface, the hooks being baited with pieces of mackerel skin, pieces of tin or any other glittering object. Pilchards are taken off the coast of Cornwall by large seine nets which are shot round the shoal from a boat which rows out from

[1] No more interesting account of the herring fishery has ever been written than that by Fulton in the *Ann. Rep. Fishery Board of Scotland for* 1899, Pt. III. pp. 242—271. Methods and legislation are alike discussed in a manner seldom seen in fishery memoirs.

the shore carrying one end of the net. Both ends of the latter are then on shore and the net is slowly dragged inwards until it touches the sea bottom and the fish are taken out from it by a smaller net. Sprats are taken in seine nets, or at times in special nets of large size. The great herring fishery is one which always goes on, for though the shoals visit the fishing grounds at different times of the year the fishing fleets follow them. The whole industry is one of considerable dimensions, for with the fleets of fishing boats go the curers, gutters and coopers, since a large proportion of the fish caught are at once cured, packed and exported. The herring fishery thus gives employment to a very large number of people, and at times the fishing ports present a very animated appearance. The mackerel fishery is of much less importance and, except in Ireland and at Milford, is carried on locally for the most part. So also with the sprat and pilchard fisheries.

Then we have a number of methods of fishing which have little general importance. Set nets, stake nets, gill nets, trammels, fishing-baulks and weirs are all in use round the British coasts. Taken collectively these methods of fishing are not unimportant, but they do not count for much so far as the supply of the great markets is concerned, and the fish caught in these ways are mostly sold locally. Set nets and gill nets are simply nets which are constructed much after the manner of a drift net but are moored near the sea bottom. Fish striking against them are meshed by the gills. So also with the trammels, though these are often buoyed like the drift nets and the fish are made to strike against them by beating the surface of the water with oars, or other things calculated to set up vibrations in the water and so frighten the fish and cause them to swim against the trammels. Stake nets are of more importance than either of the foregoing methods. They are long low walls of netting which are set on wooden stakes driven into the sands. They are arranged in straight lines at angles with each other so that the tide runs athwart them towards the corners, where there are usually pockets. Baulks are hedges of wicker work arranged on strong wooden stakes. So also are fishing weirs. Stake nets, baulks and weirs are set at the margins of channels or on sandy flats so that they are exposed at ebb tide when they are visited and fished. Usually they are owned and worked by men who are not regular fishermen but who augment incomes derived from farming or other forms of casual employment, by occasional fishing. Baulks and weirs are rather destructive "fixed engines," that is they are quite indiscriminate in their action and often take large numbers of small fish of little or no economic value. To the naturalist they are interesting apparatus, but they are generally discouraged by the fishery administrators and are maintained in virtue of some fishing right enjoyed by the land owner.

Steam trawling, steam lining and steam drifting are the organised sections of the fishing industry of these islands. The economic changes in progress at the present day apparently make for the absorption of the greater part of the fishermen into one or other of these forms of fishing. Steam trawling is carried on by limited companies, and even many of the smacks are owned likewise. So also with the steam liners and drifters. Most of the herring boats and many of the smacks are owned by the men who sail them, and often the crews are part owners of the craft they sail in and are connected together by ties of relationship. Practically all of the smaller second class boats are owned by those who work them. This is as it should be, but it may be observed that the fishermen who work the smacks, the sailing liners and drifters and the second class boats are, judged by modern commercial ideas, the unprogressive portion of the fishing population. They are, as a rule, mahogany-faced, broad-chested and heavily-built men who are usually very comfortable in their circumstances, and may be classed as a little above the ordinary man who lives in the towns and follows some artisan form of employment. Many of them own their houses and not a few the boats they work. The fishermen who work in the steamers do not possess the well-marked characteristics of the class I have just alluded to, and constitute a group of the fishing population which is fast approximating towards the class of sea-faring men who are to be found in the stoke-holes and before the mast in our modern cargo steamers. The steam trawlers are to a very great extent manned by those who have not the training of the fisherman who has been brought up under sail. Many have not the hardihood to follow the rougher though more independent life of the smacksman. Probably in a few years the steam trawlers will be manned by Chinese and Lascars. But many good things may be said of the inshore fishing population. Excluding a few who are shiftless and incompetent, who have sickened of routine work of some other kind, who work on the farms, on the docks, or wherever casual employment is to be found; and including those astute longshoremen who cater for the wants of the seaside visitors, they are a fine class of the littoral population and possess qualities which are much too valuable to the nation to be sacrificed in order that the fishing industry may be run on modern commercial lines, and that dividends may be regularly paid.

Conditions of employment vary so much round the coasts of Great Britain that it is quite impossible to notice all here. The steam trawling industry, with its command of capital, is fast becoming predominant, and there is little doubt that the other steam fishing methods will soon be similarly organised. The steam trawlers often make long voyages, and often short stays in port, for it is essential that fish should be caught. The men (apart from skippers and mates) are paid regular wages; they often have "trip-money," and sometimes a small percentage on the results of the voyage; and they are allowed the "stocker," that is the livers of the fishes and some of the less valuable species of the latter. Usually their food is provided either by the skippers or owners. It is a hard and monotonous life but does not include the same severe manual labour as on the smacks. The smacksmen are paid by wages and shares, and often entirely by the latter method. Their life is rougher than that of the men who work the steamers, but possesses many advantages. They have longer stays in port, thus many of the smacks put into harbour over the Sundays, and without necessarily expressing Sabbatarian principles one could wish that this system were a universal one. Then rough weather often means a somewhat welcome spell in harbour, and rest is thus oftener secured than in the case of the steamers. Legislation, which has abolished the infamous trade of the "cooper," has helped them in many ways; and both steam trawlers and smacksmen must have benefited greatly by the work of the Mission to Deep-Sea Fishermen. The smaller sailing boats are as a very general rule owned by those who sail them, and the conditions of employment with regard to them are of the simplest kind. They, and many of the smacksmen, sell the fish which they catch, but in the process of selling how much of the profit must find its way into the pockets of the loud-voiced, corpulent fish-buyers, those middlemen who are to be found wherever fish are landed? To those who care to investigate this economic question I recommend a comparison of the prices they pay their fishmonger with the tables of average values which are to be found in the reports of the Inspectors of Fisheries.

Finally we have the fishery for the truly benthic animals. These are shrimps, prawns, crayfishes, crabs and lobsters, with some other rarer crustacea and the mollusca—the oysters, mussels, cockles, periwinkles, whelks, &c. The crabs and lobsters are caught in pots or creels, which are baskets of various shapes and sizes, or in ironwork frames covered with strong netting. The openings of these traps are funnel-shaped orifices into which the unfortunate crustacean, which is attracted by the piece of fresh or putrid fish inside the pot, finds its way. The creels are set well below low-water mark on a rocky shore, are buoyed and moored, and are fished regularly. Prawns (*Palaemon*), crawfish (*Palinurus*), whelks (*Fusus* and *Buccinum*) are also caught by baited pots. Then we have the gatherers of shellfish. All round our coasts multitudes of men, women and children are engaged in gathering cockles, mussels and periwinkles. The cocklers[1] go down on the sands as the tide recedes, sometimes singly, sometimes in parties accompanied by a horse and cart, and they scoop out the shellfish from the sand with a kind of toasting-fork, or rake them up with a small garden rake fixed at the end of a long or short shaft. Then they wash and riddle the molluscs to eliminate those which are below the minimum legal size. They can work on the sands for a few hours only during the twelve, but these include any of the twenty-four except the darkest; are endured in all kinds of weather, and at all times in the year; often mean hardship and discomfort

[1] There is a very good account of the cockle fisheries of Morecambe Bay in the *Pall Mall Magazine* for September, 1896. If the reader can procure this it is well worth reading.

without the stimulus of danger; and are too often rewarded by a few postage stamps sent by the "commission agent." Musselers go on the scars at low water and gather the fish by hand, or rake them up; or they sail over the mussel bed at the time when the latter is just covered by the ebb-tide and rake up the molluscs by means of the "long craam," a heavy rake which is fixed at the end of a very long wooden pole. They too must observe a minimum size and also a "close season." They have most of the hardships of the cocklers and often as little remuneration. Periwinkles are simply picked up by hand, and those who eat these luxuries should reflect on the labour of picking up a hundred-weight of these shellfish, and then of carrying this weight in a bag on the back over perhaps a mile or two of sand; and further on the adequacy of a payment of perhaps five shillings for this load. Oysters are now largely imported from America, France or Holland and are relaid by the fish merchants, and fattened for the markets in selected creeks. There are still English natives to be had for dredging, but the method of importation is so well adapted for industrial organisation, and our tariff system is so convenient that the fishery for the native oyster is, in many places, no longer in existence. Shrimps are often caught by the "power net," an unwieldy semicircular net which is pushed by a man who wades in the water. He carries a basket on his back and into this he dumps the collection made by his net. Generally he looks the picture of misery when so engaged, and no doubt his appearance is a good index of the general degree of prosperity which attends this method of fishing. Shrimps too are often caught by means of "hose nets," which are tube-shaped nets kept open by means of rings, and stretched out on the sands in the tide-way. Mussels, cockles and periwinkles are sold alive, the two latter exclusively for food, but many mussels are used for bait for the long lines. A large proportion of the shrimps which are caught in England are sold fresh, but a considerable quantity are also potted by being shelled (deprived of their carapaces), boiled in butter and put into little dishes. Recently a considerable trade has sprung up with Holland in shrimps. The crustaceans are imported in sealed tins or otherwise, after pickling in a boracic acid brine, and they are potted in the same manner as English shrimps. It might be wrong to suggest

8—2

that they are sold as such, but this is not an impossible transaction. The importation of Dutch shrimps is one manifestation of the effects of our fiscal system on which Southport fishermen hold strong opinions.

The shellfish gatherers are, in the language of Mr Wells, the "Abyss" of the fishing industry. It is true that very many of these people are fairly prosperous, and apparently earn a sufficiency, but nevertheless one meets with the "dregs" of the fishing population among the people who gather mussels, cockles, and the like. They work hard, endure much physical discomfort, and are generally poorly paid for it. No one troubles about them, and restrictive legislation only inflicts hardship upon them, in many cases, without producing any obviously useful results. Some day perhaps machinery may be evolved for the collection of the shellfish of our shores. Cockles may be gathered by some ploughing of the sands, and mussels will be dredged by motor boats. The industry will then be systematised and reduced to order, but whether the shellfish gatherers will be greatly benefited may well be doubted.

PART II.
QUANTITATIVE MARINE BIOLOGY.

CHAPTER VI.

QUANTITATIVE PLANKTON INVESTIGATIONS.

MARINE biology is one of the last of the sciences that has adopted quantitative methods. This is not because the study of natural history is an unprogressive one, but just because an enormous mass of descriptive work had to be done before the exact methods of physical and mathematical science could be applied to the solution of problems of a general nature in marine biology. There are probably ten thousand species of fishes which inhabit the seas and fresh waters of the earth, and the fishes are only a group of the vertebrata, while the latter form only one of the smaller sub-kingdoms of animals. All the many hundreds of thousands of animals had to be collected, described and named; and as many as possible of their life-histories had to be investigated before it was possible to attempt any research into the inter-relationships of the different groups. Now all this work was comparative. There are no absolute standards of structure or relationship, and so it has not been possible, until very lately, to deal quantitatively with specific or morphological characters.

Then we have had fashions in zoological investigation. Prior to the publication of the *Origin of Species* the methods of natural history were, on the one hand, those of the collector and systematist, and on the other, the methods of the comparative anatomist. Collections of animals from all parts of the world were made and described, and the species so obtained were named and classified. The names of a whole army of workers will occur to the student in this connection. The older human anatomists were they who concerned themselves with the structure of animals other than man: such were John Hunter, the three Monros and many others. When however the conception of evolution, as the result of processes of natural selection, became generally accepted then the study of morphology became the fashion, and zoologists sought in the results of comparative anatomy and embryology the keys to the inter-relationships of the great groups of animals; Anton Dohrn and Huxley set this fashion. But the field of phylogenetic speculation was destined to prove a somewhat sterile one, and it has to be confessed that the study of such things as gill-slits, coelomic cavities and the like, has not afforded the results that were expected. In recent times it has become recognised that the older methods of morphology are alone inadequate for the solution of the question of the historical development of animals, and we have attacked this question on the one hand from the point of view of palaeontology, and on the other by statistical and physiological methods of inquiry.

The study of variation began with Darwin. But during the last two or three decades the variability of animal structures has been studied mathematically by Karl Pearson, Galton, Heincke, Duncker and others, and this branch of zoology, biometrics, has been developed to a very great extent. The rediscovery of Mendel's laws of inheritance led also to the development of quantitative biological methods. The results of the oceanographical voyages had the same effect. This science originally concerned itself with the determination of the areas, depths, currents, and the physics and chemistry of the oceans. But soon biological and oceanographical investigations were carried on concurrently, and it was found that much was to be learned from the consideration of the occurrence of marine organisms in relation to the physical and chemical conditions of the sea areas in which they were to be found. This led to the attempt to investigate the distribution of animals as far as possible in a quantitative manner. But there were very great difficulties in the way of investigating the distribution of the benthic population of the sea in this way, and it was thought by Hensen and his pupils that a quantitative study of the plankton would probably be attended with some measure of success.

The Hensen Methods. Hensen tells us that he sought to devise methods by which the answers to two questions might be given: (1) what quantities of living organisms in the form of plankton does the sea, in a given area, contain at a certain time? and (2) how does this quantity of plankton vary from place to place, and from time to time? He studied the plankton because it is here that the greatest mass of life is to be found, and also because it is only the plankton which is apparently capable of receiving quantitative treatment. In order to supply the answers to the questions propounded it was necessary to devise some means by which it would be possible to filter a known volume of seawater so as to remove all the organisms contained in it, and to do this expeditiously so that the experiments could be carried out over a wide sea area. It was indeed possible to make comparative estimates of the amount of plankton contained in the sea by the use of the ordinary tow-net, but even at the best such estimations can only be very imperfect ones. Plankton might be collected in quantity and the quantity might vary from place to place and from time to time, but a detailed consideration of the methods so involved will shew that the quantity of material so collected in the nets depends so much on uncontrollable circumstances that even the comparative results so obtained may be of little value from a quantitative point of view.

It is obvious that the amount of material taken by a tow-net will depend on the length of time that the net has been hauled through the water. This can, of course, be regulated very exactly. But the catch of the net depends on the amount of water which passes through the meshes, and this depends not only on the length of time that the net is hauled but also on the velocity of the haul. It is very difficult to cause a vessel of any size to travel through the water at a constant rate unless she is going fast, and it is necessary that the nets should be hauled slowly. If the vessel travels slowly less water will pass through the meshes of the net, and the latter will also sink a little below the surface of the sea and so fish in a different water layer. If she travels fast then more water will pass through the net but this will then rise to the surface, or even a little above the latter. Wherever there is a rapid tidal stream additional complications are introduced, for if the vessel travels against the stream more water will pass through the net than if she travels with the stream. Then the velocity of the tidal streams varies with the time between the neaps and the springs, and on a much indented coast its direction also varies considerably. If the net fishes at the surface the effect of the wind must also be considered, for the force of the latter is a factor of some importance in augmenting the velocity of the tidal streams.

It is theoretically possible so to regulate the speed of the vessel as to counteract the effects of the variable flow of water, but anyone with experience of this work will see how difficult this must be. One is perhaps safe in saying that it is practically impossible to avoid the effect of the tides, and to ensure that the same amount of water always passes through the tow-nets in each haul of the latter. This is one obstacle to the employment of the ordinary surface tow-net for quantitative estimates of the abundance of the plankton, but there is another which is even more inconvenient and difficult to avoid. It is well known that there is a vertical movement of the organisms composing the plankton and that this vertical migration may take place very rapidly, so that planktonic creatures present at the surface at one time of the day may be absent an hour or two afterwards. The alternation between day and night, and even between a dull and a bright sky, will produce these differences, and there is probably a more or less regular diurnal variation in the plankton of the different water strata of the sea. Changes of temperature will also lead to considerable variations, and so it is necessary that we should consider the direction of the wind if we wish to employ surface tow-nets to obtain estimates of the abundance of the plankton. In our seas a north or east wind will chill the surface and a south or west wind will warm it, and we will usually find corresponding differences in the nature of the plankton taken. Taking everything into consideration it is evident that the surface tow-net is of little use in quantitative plankton fishing.

But if instead of towing a net along the surface of the sea we sink it down to the bottom and then haul it slowly up again we get rid of most of these disturbing factors. The net can be hauled by means of a steam winch, the speed of which can be exactly regulated, and thus we ensure that the same quantity of water always passes through the net. It is possible to calculate very approximately what this volume of water is, so that we can say that the catch made was contained in a certain portion of sea. The haul can be made very quickly, and while it is being made the ship is drifting with the wind and tide so that the effects of these are eliminated. Further the net fishes through the water from the bottom up to the surface and thus it does not matter in what horizontal stratum of the sea the plankton is contained. Also the amount of plankton taken in such a net is usually small, and this is an advantage, for we can then work through it exhaustively without considerable labour. This is the principle of the Hensen method. The net is very exactly constructed; it is hauled up vertically from the bottom of the sea to the surface; and its "constants" are determined so that it is possible to calculate what volume of water passes through its meshes[1].

The Hensen net. In Fig. 24 I give a diagram of the newest form of the Hensen net. The apparatus measures about 7 feet from the opening to the extremity of the bucket. It is made of three parts, the headpiece, the net, and the bucket. The basis of the whole is an iron ring of 1 metre in diameter. Three iron stays connect this with a second iron ring which is 36 cm. in diameter: between these two rings is stretched a piece of strong white fustian, and this part, which has the shape of a truncated cone, is the headpiece.

The bucket has various forms. Sometimes it consists of a strong glass cylinder which is about 35 cm. in length and about 20 cm. in diameter. This is contained in a strong iron holder, at the upper end of which is a brass ring which is provided with three

---

[1] The net and the subsidiary apparatus are described in Hensen's first paper, "Ueber die Bestimmung des Planktons" &c., in 5 *Ber. Komm. Wiss. Untersuch. deutschen Meeres*, Berlin 1887; and in "Methodik der Untersuchungen," *Ergebnisse Plankton-Expedition*, Kiel, 1895.

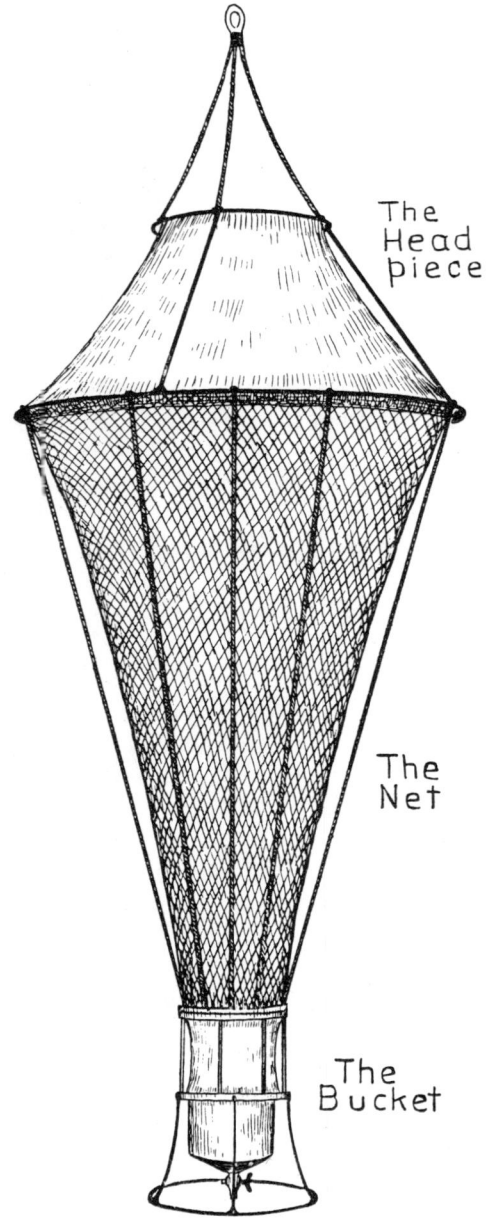

Fig 24. The vertical quantitative plankton net.

brass screws which can attach the bucket to the net. In the most elaborate nets the bucket is a cylinder made of two parts. The lower part is a copper cylinder, the bottom of which is a short cone terminating in a stop-cock; and the upper part is made of the same silk as the net itself. The metal part of the bucket is tinned on the inside and painted on the outside. Sometimes the upper silk part of the bucket is dispensed with and then this part consists only of the metal cylinder and stop-cock. In the figure the more elaborate apparatus is represented. The lower part of the guard is an iron ring which prevents the stop-cock from coming into contact with the mud at the sea bottom when the net is lowered until it touches this.

The net is made of "Müllergaze" (or botting cloth), a fabric which is used by millers for separating the various grades of flour. It is made of very fine silk thread and the meshes have various sizes. In fishing for the larger planktonic organisms the coarser cloth is used, but when it is desired to catch the very smallest things in the sea the cloth is selected which has the finest mesh obtainable. The müllergaze used in the vertical Hensen nets is usually that numbered 20 and this has, on the average, about 5900 meshes to the square centimetre. The length of side of each mesh is about 0·05 mm., and the area of the mesh is about 0·0025 sq. mm. The shape of the mesh is pentagonal and the web is so constructed that the meshes are all very similar and cannot easily be distorted.

The cloth is cut so that, when the two straight edges are sewn together, the net when opened out has the form of a long truncated cone. The wider end of this cone is 100 cm. in diameter. It is sewn to a piece of canvas and this is attached to the iron ring which forms the base of the headpiece. The smaller end of the net is 20 cm. in diameter and is also attached to a strong piece of cloth which is sewn into a piece of fustian which is folded over a brass ring which screws on to the upper part of the bucket. The net itself is not strong enough to stand the strain of hauling and the bucket is carried by another net made of strong twine, and having coarse meshes. This net is laced to the base of the headpiece and to the brass ring which forms the lower part of the net. Outside this net again there are a number of ropes which give additional security. The external net and ropes also protect the silk net from accidental injuries.

The apparatus is used in the following manner. To the upper ring of the headpiece are attached three bridles and to these is shackled the rope or steel wire by means of which the net is hauled. A light derrick is rigged out from the mast, and to the end of this is fastened a block over which the rope passes out. When the net is used from a heavy vessel, or in rough seas, the block over which the rope passes is not directly fastened to the derrick but is connected with the latter by means of an accumulator, or dynamometer, in the same manner, and for the same object, as in the case of deep-sea trawling or sounding. The rope may be marked as in the case of a deep-sea sounding line, or it may pass out over a "metre-wheel," that is a wheel provided with a counter which indicates the number of metres of rope which have passed over it, and therefore the depth to which the net has been sunk[1]. When the apparatus is to be used the ship is stopped and brought round so that if a sea is running the net may be lowered over the windward side. It is then lowered when the ship rolls over to windward, so as to avoid the possibility that it may go under the keel and in hauling foul the latter. If the ship drifts too much to leeward while the net is being lowered then the rope will lead out at an angle from the side and the net will not fish through a vertical column of water. But if a sea is running it may not be advisable to use the net at all.

The apparatus is hauled by the steam winch so that it is brought up to the surface at an even speed, and the time of hauling is taken so that the amount of water passing through the net may be calculated. The net is swung in on deck and the catch is removed. In the Hensen procedure the cock at the bottom of the bucket is opened and the catch is run out into the "Filtrator."

---

[1] On board a small ship and in shallow water the gear may be simplified. In the Lancashire fishery steamer, *John Fell*, the net is hauled from an ordinary boat's davit specially mounted for the purpose. This is swung out and in by means of guys in the ordinary manner. Both metre-wheel and accumulator are unnecessary in a small ship and in shallow water. If a steel wire rope is used this may be marked simply by painting with coloured paint at every five fathoms. When this paint is allowed to dry into the strands of the rope it is not easily rubbed off.

This is represented in Fig. 25 and consists of a metal funnel which terminates in a cylinder which is made of the same silk which is used for the net fabric. The lower end of this cylinder

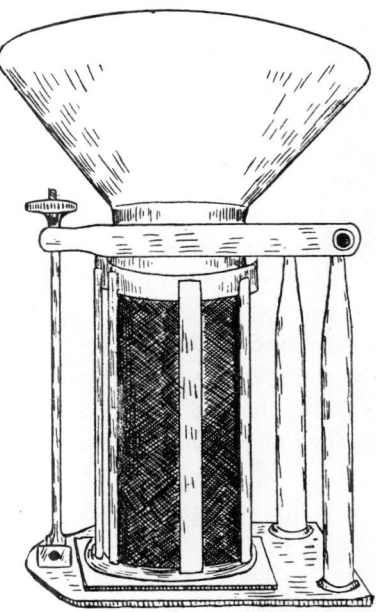

Fig. 25. The Hensen "Filtrator."

is a ground brass ring which is tightly screwed down to a glass plate from which it can be detached. The contents of the bucket are run into the funnel and the water passes through the silk side of the cylinder while the catch remains on the glass plate. After the net has been discharged it is washed on the outside with a stream of water from the hose so as to wash down into the bucket any organisms which are attached to the interior walls of the apparatus, and the bucket is again emptied into the filtrator. The frame of the filtrator is then unscrewed and the catch is removed from the glass plate, bottled and labelled[1].

[1] On board the *John Fell* the use of the filtrator has been dispensed with without detriment. When the net is hauled the contents of the bucket are simply run out into a little silk net about three inches in diameter and about six inches long. To the inside of this net a silk thread is attached so that it can conveniently be turned outside in. The catch remains at the apex of this little net.

All the water which enters the net does so through the smaller opening of the headpiece. This is much smaller than the greatest diameter of the net, and its area is from 1/28th to 1/6th of that of the total filtering surface of the net. The object of this arrangement is to allow the water which enters the net to pass through the pores of the latter with as little pressure as possible, and so avoid the destruction of delicate organisms, or the lodgement of the smaller things caught in the meshes. No filtration takes place through the fabric of the headpiece which is made of impervious fustian. The practical advantage of the headpiece is that if the net should touch the sea bottom when it has been lowered the larger iron ring rests on the bottom deposits and none of the latter enter through the opening. The headpiece has also several other objects. The net has now been hauled through a column of water the height of which is equal to the depth to which the apparatus has been lowered, and the area of cross section of which is that of the opening of the net. But it will be obvious that all of this column of water has not been filtered through the meshes of the apparatus because of the resistance offered by the fabric of the net. How much of the column of water does actually pass through the meshes and is filtered? It is obvious that only a fraction does so pass and it is next necessary to determine what is this fraction. We have to find now what is the filtration capacity of the net, and this is calculated from a knowledge of the amount of water which passes through the silk at a pressure which can be determined from the size and form of the net, and from a knowledge of the rate at which the apparatus is hauled up through the sea, a rate which can very easily be observed[1]. Once the constants of the net are known the volume of water which is filtered in each haul

The latter is then turned out, the plankton adhering to its sides, and it is taken rinsed into a quantity of preservative contained in a wide-mouthed glass ar. The last trace of plankton remaining on the silk is washed off from the latter by a fine stream of preservative directed on it from a wash-bottle. A very simple filtrator is used by the Heligoland naturalists: this consists of a brass ring three inches in diameter and three inches deep. On one end of this a piece of silk müllergaze, similar to that used for the net, is fastened by a strong rubber ring or band. The contents of the bucket are run into this apparatus; the silk is then detached and folded up and put into a bottle containing the preservative.

[1] See Appendix.

is easily calculated, and the error resulting from this estimation probably does not amount to very much. Now, at this stage of the investigation, we are able to say that the catch obtained has been contained in a water column extending from the surface to the sea bottom and of a definite cross section. This cross section is always a fraction of a square metre and it is easy to calculate the amount of plankton which has been contained in a column of sea of the depth *in situ* and of one square metre in cross section.

One condition limits the use of the apparatus just described. It is necessary that it should be used from a vessel which is stopped, and therefore it can generally be used only on vessels which are intended primarily or exclusively for scientific work at sea. But it would be a very great advantage if some form of plankton net could be devised which could be used from a steamer travelling at full speed, for in this way it would be possible to utilise the

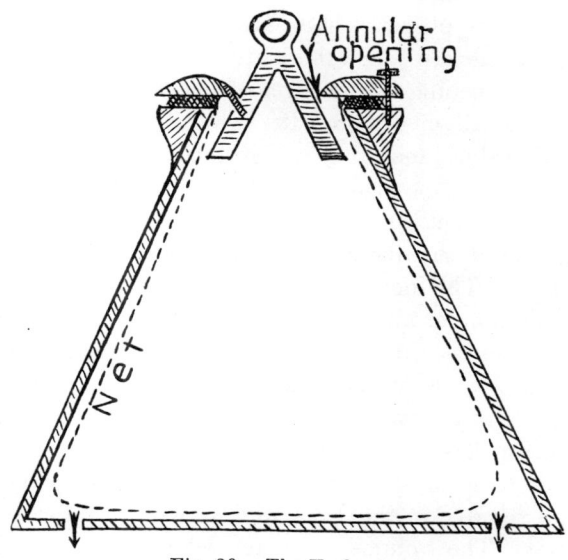

Fig. 26. The Korbnetz.

services of merchant ships traversing all seas. Several such methods of collecting plankton are in common use, but it is difficult to devise a means of quantitative fishing such as would yield the results given by the vertical nets. In 1888 Hensen devised the "**Korbnetz,**" an apparatus which could be so employed. The Korbnetz (Fig. 26) consists of a strong metal case shaped like an inverted cone, and pierced with two small holes at its base. The truncated apex of the cone has a rim, and to this rim is screwed a metal ring which is covered with fustian. A hollow metal cone is attached to the ring so as to leave an annular opening between it and the ring. The net is towed by a rope which is shackled to the cone. When it is being towed the water passes through the annular opening into the net, distending this and then passes out again into the sea through the two small holes in the base.

The Petersen-Hensen quantitative net is a very complicated apparatus and I can refer to it only very briefly. It is a cylindrical net which is thrown into deep lateral folds, which folds are produced by metal rods which are contained in the case in which the net is placed. This form of the net gives it a very large filtering surface. All the water which enters the net does so through an opening in the cover of the case, and in this opening is placed a small metal propeller which is actuated by the indraught of water, and also a revolution counter. From the number of revolutions made by the propeller the amount of water which enters the net can be calculated in an analogous manner to the estimation of the distance traversed by a vessel which tows a "patent log." The net is towed astern of the ship while the latter is travelling at an ordinary speed, and because of the small area of the opening through which the water enters there is not much pressure on the walls of the net, and the latter is not ruptured even if it is towed at a very considerable speed.

In the case of each of these apparatus the net, when the haul has been made, is removed from the case and is washed out and the catch estimated in the same way as will be described for the vertical net. The volume of water which has been filtered is found directly from the recorded revolutions of the counter, in the Petersen-Hensen net; and it is calculated in the case of the Korbnetz from the length of time during which the apparatus has been towed, the area of the annular opening, the speed of the ship, the angle of the tow-rope, and other factors. Probably the volume of the filtered water is estimated very imperfectly, and neither of the two nets just described can be said to be very successful. Further they can be worked only in a horizontal direction.

Estimation of the catch. Now we may return to our original vertical net. We have obtained a catch which consists of all the larger organisms which were present in a column of sea water of depth equal to the depth of the sea *in situ*, and of a known cross section. The nature and amount of this catch remain to be determined. In ordinary qualitative plankton work the usual procedure is to measure the volume of the catch, then pick out the larger organisms with the aid of a hand lens, and then examine a small portion of it in detail under the microscope. The organisms present are identified and their relative abundance is estimated. It may be said, for instance, that the constituents are "abundant," "frequent," "common," "rare," or "very rare," or any other scale of relative abundance may be employed. But these terms only express the relative frequency of occurrence of the organisms identified; they are purely subjective, and may differ with the personal equation of the observer; and no precise significance can be attached to them. Obviously the smaller organisms must be less correctly estimated than the larger ones[1]. They have therefore little value, and perhaps all that ought to be stated with regard to a plankton catch examined in this manner is that one or more organisms are abundant, and that others are rare. The Kiel planktologists employ much more exact methods, and a plankton catch may be examined quantitatively by one or more of four methods:

(1) by the estimation of the volume of the catch;

(2) by chemical analysis;

(3) by weighing; and

(4) by the enumeration of the individual organisms contained in it.

According to the end in view one or other of these methods is employed.

[1] Apstein, "Die Schätzungsmethode in Planktonforschung," *Wiss. Meeresunt. Kiel Komm.* Bd. VIII. Abth. Kiel, 1905.

Useful results are generally afforded by the mere estimation of the volume of the catches made: thus it is easy to compare, in a rough sort of way of course, the relative abundance of plankton in the areas investigated, merely by a comparison of the bulks of organisms contained in similar volumes of sea water. The catch is taken out of the bottle in which it is contained and is put into a measuring glass and shaken up with the preservative, preferably alcohol, since the organisms settle down more quickly in this fluid. Then after the whole has been allowed to stand for some time, at least a day or two, the level to which the catch rises in the measuring glass is read off on the graduations of the latter. This procedure requires some considerable time if a larger number of catches have to be examined, and when this is the case the separate bottles are allowed to stand for the necessary time and then the level to which the plankton rises in each is marked on the outside by means of a pencil which will write on glass. Each catch then is emptied out into another vessel and the original bottle is filled up with water to the level of the mark from a burette and the quantity of liquid discharged from the latter gives the volume of the catch. But obviously this volume is not the true volume of the organisms but is that plus the volume of the liquid which was present in the interstices between the separate organisms. Now this latter value will vary with the closeness with which the separate organisms lie together and obviously this depends on the nature of the catch. Thus copepoda lie together closely, but diatoms furnished with long spines will leave considerable interspaces between them.

If the catch is filtered through a very fine silk cloth it is obtained in a wet condition, and only a very small quantity of liquid is contained in it. Then a measuring cylinder is filled with water, or some other liquid, up to a certain mark and the wet catch is added to this liquid. The level of the latter will rise and the difference of level is the volume of the catch plus the liquid remaining in its interstices. But if minimal catches have to be measured this difference is very small, and this is the defect of the method. Again the volume of the catch in its preservative liquid may be exactly measured and then the whole is filtered. The volume of the filtrate is again measured and the difference is the volume of the catch, again plus the interstitial liquid. When very small catches have to be measured it is better to regard these as minimal ones, in much the same way as we regard the results of chemical analyses which are just on the limits of quantitative expression. All methods of plankton volume measurement are more or less unsatisfactory, and a very serious objection is the frequent presence of the larger organisms in the catch. Thus if a catch of 5 c.c. volume contains two *Pleurobrachiae* while a similar catch contains none of these organisms, it would obviously be a misleading result if the volume of the first catch included the ctenophores. The very large planktonic animals must therefore first be picked out from the catch before the volume estimation is made. Since the nets of finer Müllergaze are not adapted for catching these larger things they can be omitted without error from the catches made, or at any rate they can be estimated separately.

The mass of a plankton catch is very easily estimated. The collection is washed with fresh water until the salt has been removed, and then it is put into a weighed glass capsule and heated in a hot water oven at a temperature of 100° C. Then it is put into a desiccator and when the weight becomes constant it is recorded.

It is not possible here to consider the methods of chemical analysis. It will be sufficient to state that the estimation of carbon, hydrogen and nitrogen by the ordinary methods of organic combustion are the fundamental results. Other constituents of the catch are often estimated. Thus fat, which is abundant in copepods and other crustacea, is determined by making ether extracts. Carbohydrates, chitin and cellulose may also be determined. The inorganic constituents usually estimated are chlorine, phosphoric acid and calcium (both the latter of importance as forming the materials of the skeletons of many organisms), and silica (which is an important constituent of the skeletons of diatoms and radiolaria). The total ash which remains after ignition is also often estimated. A knowledge of the chemical composition of the plankton is, of course, of value in the discussion of many general problems, but unfortunately it cannot be obtained without destroying the catch, and when the latter has been obtained at some considerable cost, as in the case of expeditions fitted out to explore the seas in remote parts of the earth, its biological interest is often so great as to exclude the possibility of a chemical analysis.

Of all the methods of investigation of the catches made, the actual counting of the individual organisms is the most satisfactory. In this way its biological composition is known, and if necessary the chemical composition can also be determined, in an approximate manner at least, by the application of constants previously ascertained. Thus the chemical composition of the principal planktonic organisms, such as diatoms, copepods, peridinians and so on, can be determined by analyses of catches which consist predominantly of these organisms, and if the average number of each of these creatures, corresponding to a given mass, be previously determined then the chemical composition can be ascertained from the results of the enumeration. From most points of view the direct counting of the catch is the most valuable method of estimation. Now if, as often happens, a single catch may consist of millions of individuals it becomes an utterly impossible proceeding to count these, and their number has therefore to be estimated by the methods of dilution of the original catch which are employed by the bacteriologists. The original catch is mixed with a volume of water, so that it has a definite volume and then a fractional part of this diluted catch is taken and the organisms in it are identified and counted. The numbers so obtained are then simply multiplied by the number of times the fraction is contained in the whole diluted volume of the catch.

Suppose that we have a catch which on a preliminary examination is seen to consist of comparatively few fish eggs, many copepods, and a large number of the diatoms *Coscinodiscus* and *Biddulphia*. Its rough volume is, say, 4 c.c. Now it is usually an easy matter to pick out the fish eggs and count them directly. The catch is emptied out into a flat dish and the eggs are picked out with the aid of a hand lens, sorted into groups and identified and counted. Then the remains of the catch are swept up into a glass flask and diluted with water until the volume is 100 c.c. One c.c. of this mixture is taken and counted.

Now this is not so easy as it seems to be, for we are not dealing with a bacterial emulsion, or with a simple fluid mixture in which the contents are uniform in composition, but with a mixture in which the constituents to be estimated are particles in suspension and quickly settle down when it is left at rest. We cannot therefore use the ordinary pipette of the chemist to remove the fractional part. Hensen had to devise the "Stempel-Pipette" for this purpose. This instrument is shewn in Fig. 27. It consists of a strong glass flask which is closed by a bung through which passes a wide glass tube T in which there is a piston. The lower end of this piston P is a solid piece of metal which slides accurately up and down the glass tube but without much friction. Round the sides of the piston the metal has been cut away so that the volume of the cut-out portion is exactly 1 c.c., or any other convenient volume. The calibration of the piston is effected by filling the cavity with mercury and weighing the latter, and then cutting away more metal until it is found that the weight of the mercury contained in the cavity is exactly that of 1 c.c. of the latter metal. The diluted catch is now brought into the flask and the latter is filled with water up to the mark (which may indicate a volume of 100, 250, 1000 or any other number of c.c.), and the whole is thoroughly shaken up so as to produce an uniform mixture. Then before the organisms settle the piston is suddenly drawn up and the glass tube is removed from the flask.

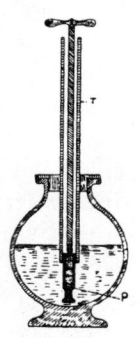

Fig. 27. The Stempel-Pipette (in section).

The cubic centimetre of liquid containing the organisms to be counted is now brought on to a glass plate which is ruled into squares, and this plate is put on the stage of a microscope which has a mechanical movement which enables each of these squares to be brought successively beneath the lens, and ensures that every square on the plate will come under observation. Square after square is then passed under the lens and the organisms on it are counted. Suppose that the catch consists of very many diatoms and of few copepods. Hensen had a box in front of him while he was examining the catch, and this box was divided into as many compartments as there were species to be counted. As each square was examined as many counters were dropped into the specific boxes as there were organisms on it. Hensen appears to have used coins as counters, but if the biologist is short of coin any other convenient objects will do—I use lead shot in pill boxes. When the plate is examined the counters in the separate compartments are then totalled up.

If the diluted catch measured 100 c.c., and if 1 c.c. of this were examined and 50 copepods were found the total number of copepods in the whole is obviously 50 × 100. Now we supposed that there were very many more diatoms in the catch than there were copepods: it may be the case then that in 1 c.c. of the catch, diluted as above, there would be too many diatoms to be easily estimated. If this is the case then 10 c.c. of the first dilution is taken and is mixed with 90 c.c. of water in another flask and 1 c.c. of this second dilution is examined as before. Suppose that this quantity contains 500 diatoms: then there were 500 × 100 × 10 in the original catch.

The enumeration of the separate organisms contained in a catch taken with a Hensen net is thus carried out by methods which are exactly those of the chemist or bacteriologist. The work is laborious but it is not at all difficult or unreliable. No doubt the limits of error are greater than in the case of the counting of bacteria contained in water, milk or air; and still more so than in the volumetric analyses of the chemists, but a consideration of the Hensen methods in greater detail than I have been able to make will shew that it is possible to describe a plankton catch quantitatively with a considerable degree of accuracy. Now we have filtered a certain amount of water from the sea in a certain small area, and we have counted the individual organisms contained in this catch. The question now arises: is this catch a fair sample of the plankton contained in the sea in the area round the spot where the net was fished?

Validity of the Hensen methods. Haeckel was the first to doubt the validity of Hensen's conclusions. In a paper of very great general interest[1] he attempted to shew that the problems which Hensen proposed to investigate were not of such a nature as

[1] *Plankton-Studien*, Jena. This memoir is written in a very attractive manner, but some of the arguments and conclusions are unreliable, and the reader should certainly also read Hensen's answer, "*Die Plankton-Expedition u. Haeckel's Darwinismus,*" Kiel u. Leipzig, 1891, before accepting all Haeckel's conclusions.

to be adapted for quantitative treatment. The general results of the Plankton-Expedition were severely criticised and, without adducing any other evidence than the general impressions of a naturalist who had fished with the older qualitative nets, he denied one of the chief conclusions of the work of the *National* Expedition, viz. that the plankton was more abundant in the colder, or temperate seas than in the warmer regions. He also laid great stress on the variability of the plankton, and emphasising the importance of swarms or streams of microscopic life in the sea (*Zoorema*, or animal streams, he called these), he contended that this apparently fortuitous distribution of the plankton militated against any attempts to estimate its abundance in a precise manner. In some respects Haeckel completely misunderstood Hensen's methods.

Then Kofoid[1] struck the next note of dissent. In a series of experiments carried out in the lakes of Illinois by means of a Hensen quantitative net he obtained results which led him to the conclusion that the catches made with this apparatus gave a very distorted picture of the actual conditions. Kofoid's first objection was that the meshes of the Müllergaze became stopped up with the smaller organisms in the water and that this closure of the pores caused the net to fish in an uncontrollable manner. His second objection was that many of the smaller organisms actually passed through the meshes and that the catch was therefore smaller than it ought to be. In order to remedy these defects of the Hensen method Kofoid proposed to substitute a pump and filter for the net of Müllergaze.

In 1899 Lohmann[2] undertook a critical investigation of the method of quantitative fishing by means of nets made of Müllergaze No. 20. Dealing with Kofoid's first objection he had no difficulty in shewing that the latter had misunderstood the nature of the method which he had criticised, and that his objection had no validity. Hensen and Apstein had shewn previously that the net had to be designed for the nature of the plankton which it was desired to catch. Generally speaking fresh water contains a greater mass of plankton in the same volume than does the water of the sea. To guard against the possible danger of a partial

---

[1] Kofoid, *Science* (N.S.), Vol. VI. p. 829.
[2] *Wiss. Untersuch. Kiel Komm.* Bd. V. Abteil. Kiel, Heft II. Kiel, 1901.

closure of the meshes of the net the filtering surface of the latter was made as large as possible when it was intended to be used in fresh water. The fresh-water net used by Apstein had an area which was 48-times that of the area of the opening through which the water entered. In such parts of the sea as were free from sediment and contained comparatively little plankton the surface of the net was only 28-times that of the opening. Now Kofoid used the smaller net, which was designed for sea fishing, and he used it in a mass of water which was rich in plankton, and over a bottom consisting of a soft sediment which was stirred up on the slightest movement. His first objection has therefore little force.

But the second objection is fully justified[1]. There are indeed many organisms which are either not taken by the tow-nets usually employed, or are taken only in very inadequate proportion. One is often struck with the absence from the catches of the tow-nets of hosts of organisms which he suspects must be present in the sea. Where, for instance, are the multitudes of the younger fish larvae which must surely be present in the sea at certain times of the year; or the naked protozoa or smaller protophyta; or the ciliated embryos of the hosts of trematodes which infest the entrails of fishes, and which pass through a free-swimming stage; or the ciliated trochospheres of molluscs which ought to be more numerous in the sea than is indicated by the catches of the tow-nets? These are either not taken by the nets, being so small that they are able to pass through the meshes of the fabric, or they are crushed out of all recognition by the pressure of water against the threads of the cloth; or they are so susceptible to injury that they are destroyed by the preservative fluid employed for the fixation of the catch. In the paper last cited Lohmann tells us that he obtained a large quantity of water from the Baltic and filtered it through the ordinary Müllergaze (No. 20, area of mesh = 0.0025 mm^2.) and then counted the organisms thus caught He then refiltered the water passing through the silk cloth by passing it through a hardened filter paper and again counted the organisms retained by this second filtration. The results were very noteworthy: most of the metazoa were retained by the

[1] It had, however, already been anticipated by the Kiel planktologists.

Müllergaze, but many invertebrate larvae and eggs were lost. The larger protozoa were retained by the net but the smaller forms were almost entirely lost. Large diatoms were caught by the net, but here again a great number of the smaller ones were passed through to be caught by the filter paper.

These results seemed to justify further investigation and so, armed with the sinews of war by the University of Kiel, Lohmann made a very searching examination of the various methods of determining the richness of the sea in planktonic life, and his paper[1] describing his results is the most important contribution to the literature of planktology which has appeared since the publication of Hensen's original memoir in 1887. It is one of those papers which stimulate thought by reason of the novelty of the results obtained, and help on new discovery by the development of a new method of research. No one method, we see, is sufficient to give us samples of the planktonic life of the sea. Just as we are compelled to employ half-a-dozen nets if we wish to obtain a true representation of the distribution of any one fish in the sea, so it is necessary to employ several different methods of research if we are to obtain a true picture of the nature and richness of planktonic life in a sea area. If we desire to estimate the abundance of the larger planktonic or pelagic animals in a sea area then it is necessary to fish with wide meshed seine nets and to pass, it may be, hundreds of cubic metres of water through the meshes of this net before the adequate sample of water has been filtered. For fish eggs we must employ an ordinary quantitative vertical net with a comparatively wide mesh, and three or four cubic metres of water will afford a sufficient sample. For copepods we may use the net of ordinary mesh and less than one cubic metre will be enough water to filter. Diatoms and protozoa can only be caught when we employ either the very finest silk cloth which can be made, or still better, the much finer filters constructed of paper or densely woven silk cloth; and for such organisms even $\frac{1}{4}$ litre of water may be a big enough sample. For bacteria quite special methods of culture must be employed, for no filter is fine

[1] Lohmann, "Neue Untersuchungen u. d. Reichtum d. Meeres an Plankton und u. d. Brauchbatkeit d. verscheidenen Fangmethoden," *Wiss. Meeresuntersuch. Kiel Komm.* Bd. VII. Abth. Kiel, 1903.

enough to retain these organisms in its meshes, and one cubic centimetre of sea water will usually be a sufficiently large sample.

Other quantitative plankton apparatus. If then the mesh of the silk cloth employed in the construction of the ordinary quantitative nets is too coarse to retain the smaller organisms of the plankton it is necessary to devise some other filtering medium. Lohmann employed a hardened filter paper which was fitted into a metal funnel in much the same way as the large corrugated filters of the chemist are used. The sample of water was run through this filter and the catch was then removed from the latter. If the sample of water to be examined is obtained from the surface of the sea then it is a matter of no difficulty to employ this apparatus; and if small samples are to be investigated then the water can easily be obtained by the use of one of the closing water-bottles used in hydrographic observations, and which can collect a water sample from any desired depth. But it is an essential of the quantitative examination of the plankton that the whole column of water from the surface to the bottom must be filtered, and this requirement introduces serious difficulties. Lohmann followed up Kofoid's suggestion of using a pump, and he devised an ingenious apparatus by means of which a fairly deep column of water could be obtained. A long hose-pipe was coiled on a windlass. One end of the pipe was open and the other was connected with a pump. The open end was then let down to the desired depth, and windlass and pump were then worked simultaneously, so that a sample of water was obtained from every level of a vertical column of known depth. This water was passed through the filter and then collected so that its volume could be measured. Thus a sea area of moderate depth can be investigated quantitatively by a method which does not appear to be open to any serious objection. The apparatus thus devised also provides a means whereby the constants of the vertical net, as calculated by Hensen's method, can be corrected, for we can filter a column of water which has, say, a cross section of two square centimetres, and since the composition of the plankton can hardly change within the limits of a column of water which is not greater than one square metre in cross section then the result of the pump method can easily be converted in that of the column of unit dimensions.

It is only with respect to the smaller organisms in the water that the ordinary nets fail, and Lohmann investigated these by means of filters made either of hardened paper or of the material known as taffeta silk. This is a fabric of thick and close strands which leave very fine interspaces between them. It was made into filters of the ordinary conical shape, or it was tightly tied on to the ends of glass tubes, and the catch was introduced into the latter and was made to pass through the cloth by gentle pressure with a little indiarubber bulb provided with an air valve, by means of which air could be forced into the tube above the level of the water to be filtered. Obviously much smaller samples are required when the smaller organisms are to be collected, and Lohmann found that even $\frac{1}{4}$ litre would yield a sufficient quantity of water to enable an estimate of the abundance of these things in the sea to be made.

Capture of plankton by appendicularians. Finally Lohmann had the happy thought of employing the houses (*Häuser*) of appendicularians to determine the contents of the sea so far as the smaller protozoa and protophyta were concerned. *Oikopleura albicans* is an appendicularian which forms a gelatinous house or case of some dimensions. This structure serves as a means of protection for the animal which lives inside it and also acts as a filtering apparatus by which the ascidian obtains its food from the plankton of the sea, and compared with it all our plankton filters and nets are very crude contrivances. The house is furnished with two openings. One is a kind of grating, the meshes of which vary in size with the size of the house. In a house which is 17 mm. long the meshes of the grating are about $127\mu \times 34.5\mu$.

This grating is the opening through which water enters into the house of the *Oikopleura*. The latter is so constructed that after entering it the water has to pass through a second, and finer filter before it flows out to the sea again through a spout. The cause of the current of water is the continual movement of the long and powerful tail of the *Oikopleura*. The animal hangs on to the internal filtering apparatus by its mouth and from time to time it sucks out the organisms which have been retained by the filter and eats them. Soon the meshes of this filter become choked up, and when this happens the house is thrown off and in a short time a new one is secreted. No one house is used for longer than about six hours.

Since the meshes of the second, internal filter are much finer than those of the outside grating it happens that organisms which pass through the grating are retained by the internal filter. Generally only such organisms as are less than 30µ in diameter pass the grating. Thus *Ceratia* and most radiolaria, and the larger diatoms such as *Chaetoceros, Rhizosolenia* and *Bacteriastrum* do not enter the house, and the only organisms that do so are the smaller diatoms, the smaller protozoa and many of the smaller protophyta, and it is only these that are to be found inside the house.

Now the area of the excurrent spout can be measured and the flow of water through it can easily be estimated. Lohmann thus found that about 162 c.c. of water passed through the house of a large *Oikopleura* in about six hours. But all the micro-plankton which was contained in this volume of water was not to be found in the internal filter at the end of this period, for obviously some of it had been eaten by the animal. Probably the catch contained in the house could be taken as that originally contained in about 100 c.c. of water. If then it was possible to kill the animal after the filtration had been carried on for some time it would be possible to estimate the quantity of micro-plankton which had been filtered from an approximately known volume of water. This was easily done, for the internal filter of the *Oikopleura* house could be dissected out and its contents examined and counted. When this was done the results were very surprising. All the organisms caught by the *Oikopleura* were such as would easily have passed through the meshes of the silk used in the Hensen nets, and many would even have passed through the much closer meshes of the taffeta silk; yet even in a volume of less than 100 c.c. of sea water there are obviously a surprising variety and abundance of planktonic life.

Thus the Müllergaze of the Hensen quantitative nets is a very imperfect filtering apparatus. If we knew nothing of the nature and abundance of the plankton in the sea except what is revealed to us by the fishing of these apparatus we should possess a very distorted picture of life in the sea. Generally speaking about half the number of the organisms which enter the mouth of the vertical net pass out again through the meshes of the silk cloth; others may be destroyed by the pressure of the water against the threads of the silk (though in a properly used quantitative net this loss is probably inappreciable in amount) and others again are destroyed by the action of the preservative liquid employed. This filtration loss too is not constant but depends on the nature of the organisms making up the catch and on the stage of development of these. Sometimes the quantitative net may only catch about 2% of the number of the organisms entering into it[1].

Nevertheless the mass of these smaller organisms which enter the net and again pass out of it is often so small that it may be neglected for many purposes. All the fish eggs, practically all the copepoda and other crustacea, and generally speaking all the adult metazoa are caught. The larger and more abundant diatoms are also caught, and so are many of the larger protozoa. *Tintinus campanula*, *Ceratium tripos*, *Peridinium divergens* are caught, but hosts of the smaller ones, such as *Skeletonema*, *Synedra* and *Prorocentrum*, are lost, and so are very many of the smaller diatoms and protophyta. Still for many of the economic problems, such as questions which can be solved by a knowledge of the distribution of the fish eggs or crustacea, the Hensen net is an efficient instrument. But we shall see later that there are other questions of theoretical importance, for the discussion of which a knowledge of the abundance of the smaller organisms which are not caught by the net cannot properly be dispensed with.

[1] The reader must refer to Lohmann's paper for tables which give the detailed comparison between the catches made by the Müllergaze, the taffeta silk and filter paper, and the houses of appendicularians. These have all been counted and the results are compared in a most careful manner.

CHAPTER VII.

THE DISTRIBUTION OF THE PLANKTON.

It is thus quite a practical proceeding so to employ quantitative fishing apparatus as to obtain all the plankton organisms which exceed a certain size, and which are contained in a column of sea water of known dimensions and position. Further, it is comparatively easy to count these organisms. The work is laborious and monotonous but presents no inherent difficulties to anyone familiar with the organisms captured. Then having made such a catch and counted the individuals contained in it we are justified in saying, for instance, that the Irish Sea 10 miles north from Llandudno contained on a certain date so many fish ova, so many Noctilucae, so many copepods and so many diatoms (to take only the more abundant organisms which would be found); and that these numbers of organisms were found in the water beneath one square metre of surface. No exception can be taken to such a statement if it is founded on the results of an investigation carefully carried out.

But should we be justified in stating that the Irish Sea 10 miles from Llandudno over an area of ten square miles contained these numbers of organisms underneath *each* square metre of surface over this whole area, and that the aggregate number of organisms could be calculated by simply multiplying the numbers of each species contained in the sample catch by the number of times that one square metre was contained in the entire area? If we say that we can do so then we assume that the plankton is uniformly distributed over the whole area referred to, and that it is generally evenly spread over the sea throughout wide areas. Now is this the case?

The Kiel school of planktologists have always maintained that wherever in a sea area the physical conditions (temperature, salinity, &c.) are uniform there the plankton is also uniform in its distribution. Plankton organisms, they say, have little powers of locomotion, certainly not such as will enable them to segregate themselves, and they are drifted about in the sea quite passively. The sea is in continual motion. The tidal streams, even at some distance from the land, have a considerable velocity and affect the water down to a fair depth. For instance in the Irish Sea the tide in the middle of the Channel will cause a floating body to be carried about 6 miles each way by the ebb and flood, and about 9 miles each way during the springs; while nearer the land the velocity of the tidal streams is greater still. Violent winds produce a considerable drift of the surface water, so much so that the effect of the tidal streams may completely be masked. The wave motion is felt down to a depth of from 30 to 40 fathoms. Storms and variable winds will thus mix up the water very thoroughly. Also sudden changes in the temperature of the air caused by changes of wind will set up convection currents which further mix the water. Because of this continual circulation of the water in a sea area the physical conditions tend to become very uniform unless there are strong currents entering it from outside. Large areas of the Clyde sea-estuary are thus practically homothermic[1], and I have seen that over wide areas of the Irish Sea the temperature may change very slowly and uniformly. If then the physical condition of the sea can be so uniform over wide areas it is not unreasonable to assume also that the plankton would be as uniform in its distribution, that is to say from place to place in a horizontal direction.

Experimental errors. It is not, of course, absolutely uniform: that is it is not possible to employ a quantitative plankton net twice in succession so as to obtain precisely similar catches in the same area at the same time. Hensen[2], on the first day of the Plankton-Expedition, made experiments with the object of testing the uniformity of the plankton. Two quantitative nets of exactly

[1] Physical conditions of the Clyde sea area; *Fauna, Flora and Geology of the Clyde Area*, British Association, Glasgow Meeting, 1901.
[2] *Ergebn. Plankton-Exped.* Bd. i.

the same construction were coupled together, lowered to the same depth and drawn up again, and seven of such double catches were made in the open sea in the North Atlantic. The catches were compared, but it was seen that no two of them were exactly alike, for when the mean of the two catches was taken as 100 it was found that the average divergence of the individual catches was 6·8 %. Also previously to these experiments Hensen[1] had made a number of experimental hauls in the Baltic by the method of making a vertical haul, emptying the net and then immediately lowering it again in the same spot to exactly the same depth. But here again the volume of the plankton taken in each of the pairs of hauls was never exactly the same, but the average divergence of each haul from the mean of the two was 15·88 % and the probable divergence 7·4 %. Now while the result of these trials was no doubt disappointing it was perhaps as we ought to have expected. It is quite impossible to construct and use two scientific instruments so that they will give exactly the same results: even with the exact and highly developed apparatus and methods of the chemists and physicists one never finds that the results of duplicate analyses or determinations are exactly alike. Experimental errors must always be encountered and one has to accept these and be as certain as possible that their magnitude is less than that of the variations which he is investigating. The vertical net is a very complex instrument and it is not possible to make two so exactly alike, or to use the same net twice in succession, so that the same result will be obtained in each case, even when we should expect that the conditions of trial were such as to lead us to expect identical results. Further the final result of the fishing operation depends on the exact estimation of the amount of water passing through the meshes of the net; and this estimation is based on the determination of the constants of the nets, and here again errors may creep into the methods.

But even when we allow for these experimental errors it is still the case that the plankton is not distributed in an uniform manner throughout the sea at any one time. We have first of all divergences from an average composition in a horizontal direction, that is the plankton is not usually the same at two parts of a sea

---

[1] *Ueber c. Bestimmung d. Planktons, loc cit.* See also Appendix.

area, say twenty miles apart, unless perhaps in the open ocean. Then the plankton is hardly ever the same at all levels of the sea and we almost invariably find that the catches obtained from different depths vary considerably. So familiar a fact is this that hosts of nets have been devised to fish only at definite depths of the sea, and any one who has used these nets will know how different the catches made at different strata may be.

Stratification of the plankton. The most careful and rigorous experiments of this kind with which I am acquainted are those which were made by Lohmann[1] in the open Mediterranean off Syracuse. Lohmann made nine catches during the period May 1—11, 1901. Each catch was made at the same time—between 7 and 8 in the morning. All the catches were made at the same place, and they were made by sinking a closing water-bottle down to a different depth on each occasion and so obtaining a water sample from the stratum to be studied. One quarter-litre of this water was filtered through thick taffeta silk, and the catch so obtained was examined alive so as to avoid the loss of organisms due to the destructive action of the preservative which would otherwise have been used. The separate individuals of the microplankton (protozoa and protophyta) in the catches were all counted. The results shewed that most of the species obtained were more abundant at a depth of about 40 metres from the surface, and that the abundance of the plankton decreased slowly from that water level up to the surface, and rapidly down to the bottom.

Heliotropic migrations. There is thus a "stratification" of the plankton. It usually varies both in nature and abundance with the stratum of sea investigated; and this vertical irregularity changes rapidly from time to time. Now in accounting for the causes of this vertical distribution we may with some probability assign two factors: (1) the variations in the amount of light falling on the surface of the sea and penetrating down to the deeper layers; and (2) variations in the temperature of the surface water. There is little doubt that the variation of the light is a

---

[1] "*Neue Untersuchungen.*" Table I. gives the results of these trials.

potent cause of the vertical irregularity of the plankton, and some proof of this may be given. If a tow-net is lowered to the bottom of the Irish Sea during the day a catch is obtained which nearly always contains *Sagitta* though this worm may be absent from the catch made at the surface at the same time. So also if a catch is made at the surface during the night this almost invariably contains numerous specimens of the same animal. Both at the sea bottom during the day and at the surface during the night, light is absent, or nearly so. If again we empty a plankton catch into a glass "crystallising dish" and stand this at a window we will always find that there will soon be a distinct segregation of the copepods: some of these will swim over to the side of the dish nearest to the source of light, while others will seek the darker parts of the vessel. The same behaviour of copepods towards light can often be observed when we put a catch of these crustaceans into a watch glass and place this on the stage of a microscope. If all light is then excluded from the catch, except that which comes up from beneath through a narrow aperture in a diaphragm, we may find that the copepods will crowd round this pencil of light. But heliotropism is even more beautifully shewn by the larvae of the nereid worm *Phyllodoce*. This animal deposits its eggs in albuminous cocoons which are laid on the shore and if one of these structures, containing motile trochophores, be cut open in a glass dish containing sea water the larvae will at once swim across to the light and form a dense layer on the sides of the glass nearest to the source of light.

Probably the majority of the organisms forming the plankton are affected to some extent by changes in the intensity of light, and although no very systematic or exhaustive observations have been made on this subject it is nevertheless the case that considerable vertical migrations are the results of this cause. We should expect to find, and indeed do find, that there is a diurnal vertical movement of the plankton accompanying the alternations of day and night, and possibly there is also a seasonal variation in the nature and abundance of the surface plankton as the result of the yearly variation in the amount of light which falls on the surface of the sea. Finally one may conjecture that the difference between the plankton at the surface on a bright day, and that obtained on a day when the sky is overcast is also to be attributed to the heliotropism of the planktonic organisms[1].

Temperature changes in the surface layers of the sea produce vertical plankton movements either because the organisms are affected by these *per se*, or because convection currents are set up, or because changes in the viscosity of the sea water are the results of temperature changes. Apart altogether from general seasonal changes, which affect large sea areas simultaneously, we find that comparatively rapid changes in the temperature of restricted areas of the sea are the result of local changes in the direction of the wind, or are simply diurnal changes. The daily range of temperature at the surface is from 1 to 2° C., the surface being hottest at about 4 p.m. and coldest at about sunrise. Much more considerable changes are produced by changes in the direction of the wind. In the Irish Sea an east wind lasting several days will cool the surface water several degrees, and a south to south-west wind will as rapidly warm it. The change produced by the wind does not only affect the surface, but will also affect the lower layers, for the superficial film of water is rapidly removed by the force of the wind and a new layer is exposed. If the wind is blowing offshore the surface water is drifted away from the land and deep water flows inshore as an undercurrent to take its place. Conversely if the wind is blowing on-shore, surface water is banked up at the beach and then flows away as an undercurrent. If the wind lasts for some time then all the water in a moderately deep sea will by-and-by be cooled or warmed as the case may be. Ultimately then a uniform mixture is produced, but for a short time after the change of wind the temperature difference between the surface and

[1] Gough (*Fishery and Hydrographic Investigations in the North Sea and Adjacent Waters*, Report 2, *Southern Area*, Cd. 2670, 1905, p. 345) has made some instructive observations of these movements due to light changes and temperature changes. He made a number of hauls in the English Channel during the 24 hours by vertical and horizontal nets. The samples were taken every two hours. The composition of the hauls made by the vertical nets did not vary much during the whole period of day and night. But the horizontal hauls shewed considerable differences. *Euchilota pilosella*, *Obelia* sp. and *Tomopteris helgolandica* came to the surface only at night. *Calanus finmarchicus* swarmed at the surface during the hottest part of the day, but at night it dispersed through all layers of the sea. Next day it did not come to the surface, but the weather had changed, the sky was overcast and a strong wind was blowing.

lower layers will be accentuated. Chilling of the surface layers of the sea will also produce convection currents when the force of the wind is not strong enough to set up very decided surface drifts. The cold water at the surface becomes denser and sinks down and warmer water from underneath rises up to replace it. No doubt the cooling of the surface water must affect the organisms living there, merely because many of them must be susceptible to changes in temperature, but the main effect is probably the establishment of convection currents, and it is these which lead to the vertical movement of the plankton, and ultimately to a more or less complete mixture.

Viscosity changes. Ostwald[1] has made the interesting suggestion that vertical movements of the plankton are due to changes in the viscosity of the sea water. Small organisms, such as those of the plankton, are particles in the physical sense and behave as such. When the temperature increases the viscosity of the water decreases and then the plankton organisms sink towards the bottom. Diffusion currents will bring them up again. Ostwald traces the annual spring maximum of plankton to this cause and attempts to shew that the daily maximum is also due to the changes in the viscosity of the sea water. Such movements are quite conceivable and one may accept them as one of the causes of the observed irregularity in the vertical distribution of the plankton, but before we can attribute the annual maxima of abundance to these causes much experimental work remains to be done.

Inshore and offshore plankton. Now such irregularity in the vertical distribution of the plankton affects only the hauls made by nets which are dragged through the water in a horizontal direction, or the catches of nets which fish only in definite layers or through restricted strata, and it can largely be avoided by using vertical nets which are hauled up through the water from the bottom to the surface and so fish equally through all layers of the water[2]. If then we find that vertical hauls made

[1] " Zur Theorie des Planktons," *Biologisches Centralblatt*, Bd. xxii. Nr. 19.

[2] Not entirely avoided because the migrations of the plankton due to the causes mentioned may not be entirely vertical. Where there are strong tidal streams or a surface drift due to wind, oblique movements of the plankton must be produced at

at about the same time in adjacent parts of the sea give different results and if these differences are greater than can be explained by experimental errors then we can only attribute them to the irregular manner in which the plankton is distributed from place to place in the sea. And just as one is inclined to minimise or magnify the results of quantitative plankton work, so one is inclined to lay stress on, or partially to ignore these irregularities of distribution. In considering them one ought, however, to remember that the plankton is far more irregular in its distribution near the shore than further out to sea.

Many causes combine to render the plankton very irregular in its distribution near the land (say within ten to twenty miles from the shore). The tidal streams are far more rapid there than out at sea. In estuaries and bays they run with great force, curving in round the latter and forming "races" round headlands. Temperature changes too are far more rapid and extreme near the shore than further out. A wide area of sandy shore may be laid bare by the tide and rapidly cooled in the winter, or as rapidly heated in the summer, and when next the flood tide covers this area the water becomes cooled or heated as the case may be; and thus irregularly heated areas of water are brought into existence. Soon the movements of the sea smooth down these irregularities, but generally speaking as we sail out from the land in the summer the sea becomes colder, while in the winter months it becomes warmer. The salinity of the sea is always lower near the land because of the fresh water brought down by rivers. But land water always brings down dissolved matter, such as salts of nitrous and nitric acid and ammonia, and other substances which are the ultimate food-stuffs for the plankton. Thus a river grossly polluted with sewage matters will have a most remarkable effect in increasing the planktonic and benthic life in the neighbourhood of its outlet. Further the richest benthic fauna is to be found in relatively shallow water, and it will at once be seen that accumulations of such animals as molluscs, crabs, cirripedes, echinoderms and the like must add considerably to the

the time of the change of the surface movements. This is perhaps the main cause of the irregularity of distribution of the plankton in an area of uniform physical conditions.

plankton of the sea above them by the addition of swarms of larvae emitted at their spawning periods. I have seen practically pure collections of larval creatures taken in the tow-net (spatangid plutei, cirripede nauplii and cypres, crab zoeae and megalopae and so on) in inshore waters just as the net happened to encounter a swarm of these larvae resulting from the simultaneous hatching of great numbers of eggs just previous to the time of collection. All these causes must lead to irregularity in the distribution of the inshore plankton: currents by concentrating larval swarms; temperature and salinity changes *per se*—how exactly we do not know; the pollution of the sea by rivers and sewers adding food-stuffs; and the rich benthic fauna by adding spawn and larvae. Out at sea the continual movement of the water smooths down these irregularities of distribution, but in inshore waters we expect to find, and do find, a plankton which is often highly variable from place to place, and from time to time.

Thus the physical conditions in a sea area within twenty miles of the land are very different from those without this limit. In the latter case the conditions are much more uniform and so we have a more uniform plankton. Still further away from the land, in an oceanic area we find still greater uniformity[1]. The winds are more constant; and as they are unaffected by immediate proximity of land we find that the air temperature is more constant. The sea there receives no contribution from the land that can be recognised. Tidal streams are absent or are slight. Therefore because the physical conditions are so uniform we find that the irregularities of the plankton which are so apparent near the land do not exist, and the unequal catches which are made by a vertical plankton net in such a situation are probably due almost entirely to the errors of experiment.

Plankton types. Nevertheless in a wide oceanic area such as the North Atlantic we do find that there is a gradual change in the nature of the plankton from place to place. This change is due to the existence of physical changes in the sea such as are caused by ocean currents Even if the Atlantic were a streamless ocean we should still find a gradual change due to the change of climate

[1] See Appendix.

with varying latitude. If this were so then the plankton would change uniformly from north to south. But the floating pelagic life of the Atlantic is distributed much more irregularly than this, and its nature in different parts of the whole area is determined by the existence of the two great current systems—the Equatorial and Polar Stream circulations.

P. T. Cleve[1] was the first to attempt to give a general account of the distribution of the plankton over the whole area of the North-Western Ocean. Dealing mainly with the diatom forms as constituting the most characteristic organisms of the pelagic life of the sea he shewed that there were six main types of oceanic plankton in this area. Including the Peridinians and Oscillatoria these were:

Tripos-Plankton: diatoms are relatively scarce in this type, the characteristic organisms being Peridinians and micro-crustacea. It is a form of plankton which is characteristic of the sea on the north-east coast of Scotland, round the Shetlands, and off the coasts of Norway.

Styli-Plankton: diatoms are abundant and the principal species are *Rhizosolenia styliformis* and *R. alata*. *Chaetoceros lorenzianus* (in the south) and *C. gracillima* (in the north) are also present. Styli-plankton is characteristic of western Europe down to N.E. off Bermuda.

Chaeto-Plankton: the characteristic diatoms are *Chaetoceros borealis*, *C. decipiens*, and *C. constrictus*. It is a N. Atlantic plankton.

Desmo-Plankton: the principal organism is the alga *Trichodesmium*. This is the prevalent plankton of the Antilles and Brazil currents.

Tricho-Plankton: the principal diatoms are *Thalassiothrix longissima* and *Rhizosolenia semispina*. This is an Arctic plankton and is found in the sea round Iceland, in the Irminger Sea and in Davis Straits.

Sira-Plankton: this type is characterised by the prevalence of the diatom *Thalassiosira nordenskiöldii*. It is found in the Arctic Ocean.

[1] "*Phyto-Plankton of the Atlantic and its Tributaries*," **Upsala, 1897.**

Now if we study Cleve's chart of the distribution of these plankton types in the North-Western Ocean during the summer we will see that they are to be found over regions which have very direct relationships to the distribution of the waters arising out of the two great water circulations to which I have referred. We will take them in the order mentioned.

Tripos-Plankton (with several "neritic" types) is found in the summer round the coast of Norway, in the North Sea, in the Skagerak, to the south of Ireland, in the Bristol Channel and in the Irish Sea.

Styli-Plankton is found in the Atlantic between latitudes 40° N. and 45° N., in the European Stream area, in the Bay of Biscay and in the English Channel.

Chaeto-Plankton is found also in the European Stream area between the areas of *Tricho-* and *Styli-*Planktons.

Desmo-Plankton is found in the west Atlantic south of latitude 40°.

Tricho-Plankton is present to the south and east of Newfoundland, Greenland, Iceland, and in Davis Straits.

Sira-Plankton is found in the sea near north Iceland, in Davis Straits, off Labrador and Newfoundland.

Now the reader must remember that the chart (which I reproduce as Fig. 28) is only, as Cleve says, "a first essay in a schematic plankton chart of the Atlantic, summer state." When the reader considers the difficulty of obtaining simultaneous plankton observations over such a wide area as the North Atlantic, and further that such collections must be accompanied by observations of the temperature and salinity of the sea, he will see that any single attempt at a schematic representation of the distribution of the plankton over such a wide area must be imperfect and that many such series of investigations must be made before we can be sure that we have a normal sketch of the conditions of the area investigated.

Nevertheless we see in this chart a rough approximation of the areas covered with plankton of distinctive types, with the areas over which the sea water has common physical characters.

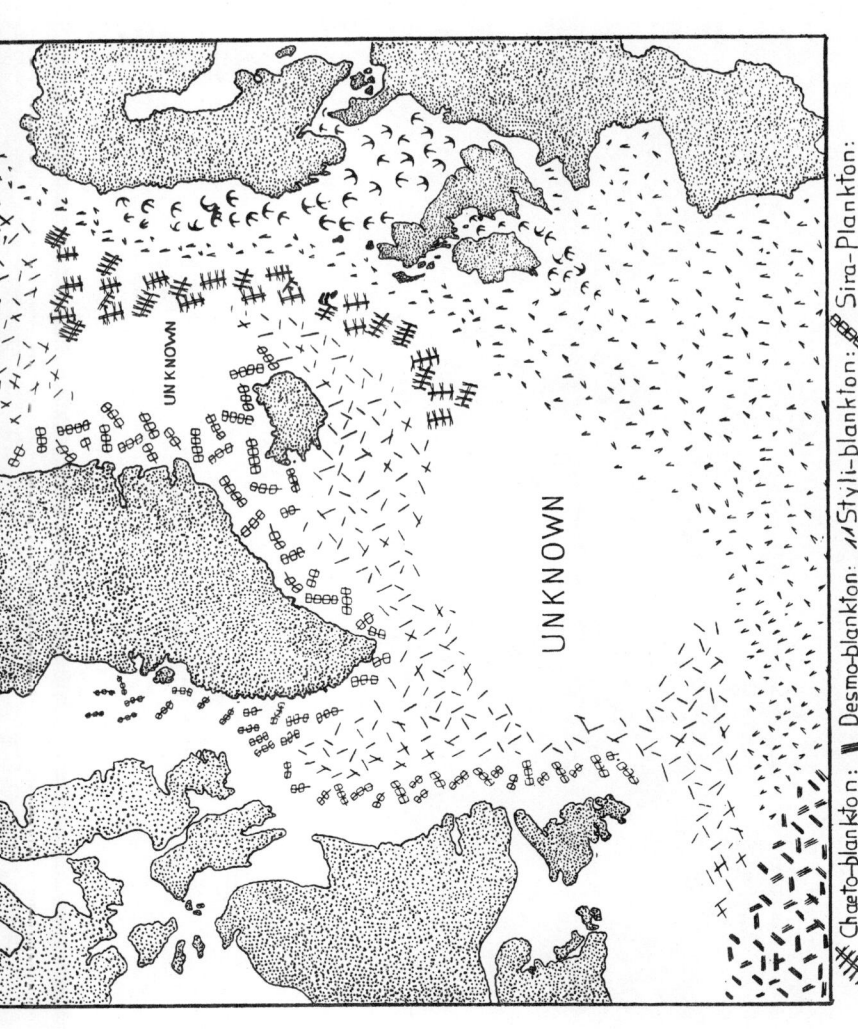

Fig. 28. Distribution of the chief plankton types in the North Atlantic during the summer. (From Cleve, "*Phyto-plankton of the Atlantic and its Tributaries.*")

Remember that there are two main sources of water in the North-Western Ocean: (1) water derived from the two great Equatorial streams, and (2) water of Arctic origin. Then we have (3) mixed water at the places where the offshoots from these two streams meet; and (4) near the land we find littoral water. Now it is easy to see from the chart that the area covered with Styli- and Chaeto-plankton corresponds roughly with the area covered with the Gulf Stream drift in the summer, that is at a time when the extension of this water towards the north-west coasts of Europe is at its minimum. Now Styli- and Chaeto-plankton are the types of the plankton of the sub-tropical North Atlantic and a detailed consideration of the species of the latter will shew that there are many which are common to it and the plankton of the Gulf of Guinea, and even of the Indian Ocean. These are carried far from the regions in which their distribution is normal and thus the occurrence of sub-tropical forms in the cold water of the Faeroe channel—a peculiarity of distribution which is so puzzling to many zoologists—receives a very simple explanation.

Then note that the water off the coasts of Greenland and Iceland, and in Davis Straits, contains Sira- and Tricho-plankton. This is the type of plankton which has been described by several expeditions as occurring abundantly in the Arctic seas. When we remember that the southerly flowing currents of Arctic water pass to the east and west of Iceland, and down from Davis Straits, and that a well-known stream of cold water—the Labrador current—flows to the south along the coasts of North America, then it is easy to see that the distribution of these plankton types is just that of the southerly flowing Arctic water which enters the North Atlantic. To the south of Greenland the two plankton types mix and this mixed pelagic life is that which is characteristic of the Labrador stream as it flows along the shores of Newfoundland.

Two other plankton types are shewn on the chart. The Desmo-plankton contains as its principal form the alga *Trichodesmium*, a plankton form which we recognise as being characteristic of tropical waters. It is found in the Atlantic in the Antilles current, and also in the Brazil current—both offshoots from the South Equatorial stream. Tripos-plankton is the oceanic type which is characteristic of the north-eastern part of the Atlantic and we see that it fills up the North Sea, the Irish Sea, extends into the Skagerak and forms a narrow edging along the Norwegian coasts. In Cleve's chart the Tripos type also includes those forms of plankton which are characteristic of the shallow seas.

Just as a number of oceanic types of plankton can be recognised so types of plankton which are characteristic of the shallower seas can also be distinguished. By shallow seas I do not mean the restricted and highly variable inshore waters within 10 to 20 miles from the land but the wider expanses of water such as the North Sea, the central parts of the Irish Sea and St George's Channel and so on. Here we find a more uniform plankton than that which is to be seen inshore, and which can easily be divided into several main types. Thus Cleve recognises the following shallow water or neritic plankton types:

Didymus-plankton, a category which includes a number of diatom and peridinian forms, and which is named from the diatom *Chaetoceros didymus*, an abundant species. It is a southern North Sea type.

Northern-Neritic-Plankton. The diatoms *Leptocylindrius danicus*, *Skeletonema costatum*, *Chaetoceros laciniosus*, and *Lauderia annulata* seem to be the most characteristic forms. It belongs to the northern shallow water area.

Arctic-Neritic-Plankton is characteristic of the shallow-water Arctic seas. It is highly complex and its constituents cannot easily be summarised.

Concinnus-Plankton is a characteristic North and Irish Sea plankton which appears in the early spring. It contains the familiar diatom *Coscinodiscus concinnus* as its typical constituent.

Halosphaera-Plankton. Now and then the shallow water plankton all round the British seas contains the "unicellular alga" *Halosphaera viridis* as its principal form. In the Irish Sea it occurs chiefly in the early summer.

Now the above names have been applied to certain plankton collections because the latter contain one form which is usually so abundant as to give the gatherings a certain individuality. Associated with these there are, of course, a number of other species. The neritic plankton-types are so called because they seem to be in some way dependent on the sea bottom in shallow water seas. In the latter, just as in the open ocean, we may have well-marked plankton-types which have often well-defined limits of distribution. If the reader will refer to Fig. 29, which is a reproduction of Cleve's chart of the distribution of the plankton of the North Sea in January 1897, he will see that this was the case then. The central part of the North Sea was covered with Tripos-plankton. Near the land on both the British and continental sides was a broad edging of Concinnus-plankton; and between this and the area of Tripos-plankton was a band of Halosphaera-plankton. Then in the Skagerak Arctic plankton seems to have prevailed for that area was covered with the Sira- and Tricho-types.

We have considered these types as if they were always well-defined and easily recognisable plankton groups, but a little consideration will shew that this is not always the case. If the nature of the plankton is intimately dependent on the physical conditions of the water in which it occurs then we should expect to find that wherever there is a mixture of the water coming from two different currents there we should have a mixture of the plankton which normally inhabits those streams. Thus we have mixed plankton-types in which species of widely different sources of origin are to be found. Then where an oceanic current, such as the European Stream, or the Labrador current, flows through an extensive sea-area, it will encounter very different climatic conditions as the latitude changes, and so it may happen that a gradual modification of the type of plankton, which was originally characteristic of the stream, takes place with the change of latitude. Thus we find that though the water of the European stream from the Azores to the Faeroe Channel, and then to Cape North, is always the same and is unmixed usually with water from any other source yet there is a gradual reduction of temperature, and other meteorological conditions change also. Therefore the organisms of the plankton, which were originally sub-tropical ones, find a gradual change of life conditions and so those which are most susceptible to change die out as the stream passes to the extreme north-east. So also with the southerly flowing Arctic stream which passes into warmer latitudes.

As the oceanic currents approach the land their characteristic plankton changes, not only because the water of the currents becomes mixed with water derived from the land, and many organisms which are susceptible to changes in the salinity of the water in which they live die out; but also because a new plankton—a neritic type—which is more adapted for life in shallow water already exists there. In the struggle for existence, that is for the always scanty food-stuffs which are dissolved in the sea, this neritic plankton, which is better adapted for the conditions of life in shallow water, predominates, to the partial or entire exclusion of the oceanic forms.

Thus owing to the mixture of the different currents with each other and with the coastal waters, all kinds of mixtures of the plankton may be produced, so that sub-types can often be recognised. Seasonal changes in the life of the great currents may be observed, and there are also seasonal changes in the composition of the neritic plankton. Add to these causes, which produce mixed plankton types, those periodic seasonal changes in the direction and time of maximum flow of the currents, and also the unperiodic changes to which I have already referred, in the case of the flow of the Atlantic stream into North European seas, and we will see that changes of the greatest complexity may take place in the composition of the plankton inhabiting any one sea area.

An endemic plankton which neither receives immigrant forms from without the area in which it is found, nor sends emigrant species to the outside of its own area, is seldom found. Only in the "Halistatic," or streamless sea areas, do we find such a plankton. Such halistatic sea areas are found wherever, as in the case of the Sargasso Sea, we have a true circulation round a centre, so that in the middle of this swirl of water there is a part which does not move; or in a lake or an almost land-locked sea; or in such places, as in the Irish Sea between the Isle of Man and Ireland, where there is no definite current, and towards which the tides set from opposite directions so that they interfere with each other and so produce a streamless region. In such areas the physical conditions change only with the seasons, and the plankton changes are also so determined.

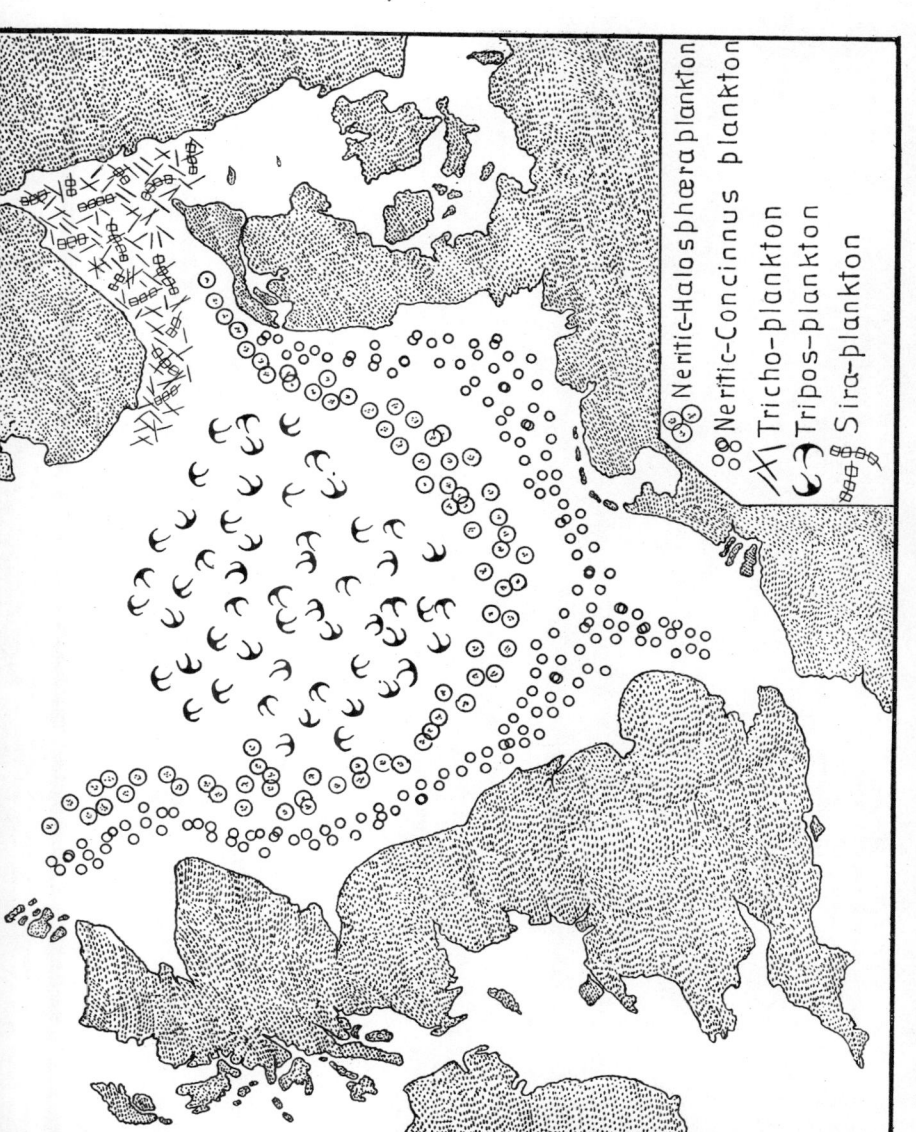

Fig. 29. Distribution of the chief plankton types in the North Sea in January, 1897. (From Cleve, "*Phyto-plankton of the Atlantic and its Tributaries.*")

CHAPTER VIII.

A CENSUS OF THE SEA.

I HAVE discussed the question of the distribution of the plankton before quoting any general statements regarding the abundance of life in the sea. Obviously the value of these statements will depend on whether or not a sample collection made with a quantitative net in a certain part of the sea represents also the contents of the sea a mile or two further off. This is just the difficulty that is encountered in all investigations that depend on sample observations. Weather forecasts depend on the determinations of the positions and movements of atmospheric pressure disturbances, and these latter are based on the readings of a small number of barometers, scattered over a wide area. So also the calculation of the average annual rainfall of a county is made from the readings of a few rain-gauges situated here and there over the district, and the amount of rain at those places where there are no instruments may vary, more or less, from that recorded at the observing stations. Our conclusions regarding the geographical distribution of animals on the surface of the earth are based on the results of sample collections, and the latter are necessarily few in number, and are often imperfect. There is always a certain degree of uncertainty in results which are based on the study of samples. It must be assumed that the samples represent the conditions that obtain at other intermediate places, and this is not necessarily the case.

The validity of all conclusions as to the general abundance of microscopic life in the sea depends on the truth of the postulate, that wherever in the sea the physical conditions are uniform, there also the composition and abundance of the plankton is uniform.

If this is not the case, if the plankton is distributed throughout the sea in a haphazard or fortuitous manner, then we cannot make any general statements about its abundance, and the investigation of marine microscopic life, apart, of course, from the collection of material for speciographic or morphological study, is futile, a proceeding "without rhyme or reason." But I do not think that the student of marine biology will readily come to this conclusion. It is true that there are multitudinous factors in operation which cause the plankton near the shore to be highly irregular in its distribution; and it is also true that little reliance can be placed on the collections made by an ordinary surface tow-net, as representing the general contents of the sea in microscopic life. And we know also that the quantitative plankton net is a somewhat imperfect instrument. But apart from a comparatively narrow margin of coastal waters, the physical condition of the sea may be the same over wide areas, and when we make hauls with a vertical plankton net in such offshore waters we find that the composition and abundance of microscopic life is usually pretty much the same over areas of sea many square miles in extent[1]. There is, in fact, a certain order in the distribution of plankton, just as there is a certain degree of constancy and regularity in the occurrence and density of the bottom living animals, or the migratory fishes. If we find that the nature and amount of the catches made by the quantitative nets are, roughly speaking, similar over wide areas of sea, then it is quite a valid proceeding to make statements of the average numbers of the various organisms which are present in the sea at certain times in the year, and per unit volume of water. But the reader must understand such results are only approximate ones, and that errors arise not only from the imperfection of the methods, but also from the lack of uniformity in the distribution of the plankton.

There is a certain mass of life always present in the sea (1) in the form of microscopic and macroscopic organisms drifting passively with the movements of the water, this is the plankton; (2) in the form of actively swimming fishes and other animals (nekton); and (3) as bottom living plants and animals (benthos). It is (theoretically) possible to estimate the absolute number of

[1] See Appendix.

organisms of the first category present per unit volume of seawater in the manner already described. It is also possible to estimate the numbers of those fishes producing pelagic ova which inhabit the same sea area; and one can also attempt to estimate the numbers of the bottom living invertebrates which likewise produce pelagic eggs and larvae. But there is no means of estimating the numbers of the fishes and other large animals which produce eggs or young which do not appear in the plankton. One can trawl or dredge for these organisms and calculate the numbers caught per unit area of sea bottom swept by the fishing instrument; but it is impossible to say with certainty what proportion of the animals living on the bottom were caught by the trawl and what proportion escaped capture. Such fishing experiments only tell us imperfectly what is the relative density of the things captured in different parts of the sea; and what changes are taking place from season to season. The material does not, in fact, exist for the construction of a "census" of a sea area: and I only quote such results as will give the reader some idea of the extent to which the sea is populated by some predominant marine organisms.

Density of plankton in the North Sea. In 1903, and succeeding years, the German Fishery Research steamer *Poseidon* made a number of hauls with the Hensen quantitative vertical plankton net in the North Sea on the eastern side and in the Baltic, and the results of these experiments have been published[1]. A cruise was made every three months and on each of these quarterly cruises a number of "stations" were visited. The number of stations varies from three to fifteen. The procedure at each station is to lower the net down to the sea bottom, and then haul it up slowly to the surface. The catch made is then preserved and subsequently counted. The column of sea water thus filtered is only a small fraction of a square metre in cross section, but the results are converted into numbers of organisms captured per column of water of one square metre in sectional area. This does not involve any error of moment, as the distribution of a

---

[1] Apstein, "Plankton in Nord- u. Ostsee auf den deutschen Terminfahrten," *Wiss. Meeresuntersuch. Kiel Komm.* Bd. IX. Abth. Kiel, 1906.

vertical sample of plankton can hardly be expected to vary within the limits of such a column. The following table gives the results of four hauls taken at stations 2 and 3 in the North Sea[1]. I have simplified the table from those given by Apstein.

Principal Plankton present in the North Sea (German Stations 2 and 3) in a column of water 1 square metre in cross section.

Organisms	Estimated Nos. of individuals			
	February	May	August	November
Halosphaera....................	8,000	—	—	—
Diatoms	3,440,000	85,400	61,800	3,091,200
Protozoa	377,600	472,000	11,744,000	372,000
Coelenterates	80	2,720	400	80
Planariae	—	80	—	160
Actinotrocha	80	—	—	240
Polychaete larvae	80	1,600	160	400
Sagitta	5,360	—	6,960	3,040
Copepods	96,160	489,720	1,344,400	198,000
Cladocera	—	—	960	—
Pteropods.....................	4,160	160	144,000	36,000
Lamellibranch larvae......	5,600	40,000	40,000	32,000
Gasteropod larvae	—	—	—	80
Zoeae	—	80	—	—
Fish eggs	—	80	—	—
Oikopleura	—	—	32,800	2,400

Now these figures represent the results of four typical plankton catches and will give the reader an idea of the nature and abundance of microscopic life in the sea—as revealed by the employment of silk bolting cloth, No. 20, as a filtering medium. Many different genera and species of diatoms, protozoa (chiefly peridinians), copepods, &c., were obtained, but I have grouped all these together and the reader who is curious will find the detailed lists in Apstein's Tables. It will be seen that the diatoms, peridinians and copepods are the most abundant organisms in the sea and that one or more millions of some of these things may be contained in a column of sea-water about 40 metres, or less, in height, and one

[1] Lat. 55° 40′ N., Long. 3° 30′ E. is the mean position of the stations. They are situated on the N.E. of the Dogger Bank.

square metre in cross area. The worm *Sagitta*, and some kind of coelenterate (usually the ctenophore *Pleurobrachia*) were always present, but the larvae of the metazoan animals occurred less constantly. The table shews that pretty much the same groups of organisms occurred during each of the four months; and that they were present in much the same relative proportions. But there is a seasonal change in progress during the year, and this is better seen when the detailed tables are consulted.

If we consider all the stations investigated during each cruise we may obtain an *average* catch, which is in some respects more instructive than the individual catches. It would require too much space to give these average catches for all the organisms considered, so I have grouped together all the species of each of the common diatoms *Biddulphia, Rhizosolenia,* and *Chaetoceros,* of the peridinian *Ceratium,* and of all the copepods, and give these average catches in the following table.

Average numbers of five kinds of planktonic organisms contained in a column of water of 1 square metre of cross area. Middle of North Sea.

	February, 3 hauls	May, 15 hauls	August, 13 hauls	November, 9 hauls
Biddulphia	128,000	9,070	1230	10,461,260
Coscinodiscus	149,000	131,200	26,600	415,000
Chaetoceros	1,303,200	13,300,000	2,660,000	140,409,000
Ceratium	1,402,000	2,973,400	8,413,600	12,172,000
Copepods	215,440	560,000	1,028,000	497,800

Variation of density of plankton with locality. In the Appendix I give the numbers of organisms belonging to each of the above four genera, and the numbers of all the copepods, taken in each of the individual hauls. Now the reader should consult these in order to see whether or not the above averages have any value as representing the contents of the whole sea area studied. The distribution of the copepods is pretty uniform, and I think that the average catch possesses some degree of reliability. The occurrence of the peridinian *Ceratium* is not so irregular as to preclude the possibility of deducing an average. But *Biddulphia* is very irregular and the average catch probably does not possess much value. There was always a large number of the diatom *Chaetoceros* present, but possibly a considerable proportion of the individuals of some of these species which entered the mouth of the net escaped again through the pores. The averages for *Chaetoceros*, then, are probably minimum ones.

Now, considering Hensen's much-quoted postulate, we expect to find that the plankton is uniformly distributed in a sea where the salinity, temperature, &c., are also uniform. If we consider (say) the distribution of the diatom *Coscinodiscus concinnus*, at the fifteen stations investigated in August we will find it to be extremely irregular. But before we conclude that the postulate of uniformity is fallacious we must determine whether or not physical uniformity prevailed over the area included by these fifteen August stations. If the reader will refer to Cleve's plankton chart (Fig. 29) he will see that this diatom is the predominant organism in a plankton-type which is restricted to a well-marked zone of the North Sea. Here then is the explanation of the lack of uniformity in the catches made by the *Poseidon*: they were made in sea regions characterized by different physical conditions. In an ideal scheme of plankton investigation there would be half a dozen steamers making simultaneous quantitative collections during three weeks of each of the four sample months of the year. These vessels would cover the North Sea with a network of lines of observation and both plankton and hydrographic work would be done at each station. Then we should find that there were a number of zones each characterized by water of a certain mean salinity and temperature. Over each zone we should expect a tolerably uniform plankton content. Then all the hauls taken on each zone would be grouped together and the average of these would be taken. We should thus find, not that there was an average plankton content for the whole North Sea, but that there were a number of sub-areas, and that over the whole extent of each of these sub-areas, the plankton would be so similar that it would be possible to deduce an average value for its composition and density. Further, by the application of the principles outlined in the Appendix we should be able to state how much this average deviated from the probable value, and so to assign limits of error to our estimated average values.

Such a scheme of investigation is perhaps a Utopian one, and in the meantime one must take the results as they are. Their perfection is of course highly desirable, but until this is attained one must not be precluded from extracting what value he can from admittedly imperfect data.

Diatoms. These are the most abundant organisms in the plankton[1]. The average catches which I have quoted shew that there were, in February 1903, over one million of *Chaetoceros*, and about ¼ million of *Coscinodiscus* and *Biddulphia* underneath each square metre of surface of considerable areas of the North Sea on the continental side. Now the *Chaetoceros*, and some of the smaller *Coscinodiscus* species, were probably underestimated; and other genera of diatoms were also present. But enormous as these numbers are, they do not represent the abundance in which the diatoms appeared at times in Northern seas. Brandt[2] quotes a haul made in the Bay of Kiel in which there were 3173 millions of diatoms (chiefly *Chaetoceros*). This mass of life was taken by a net which was lowered to twenty metres beneath the surface and then drawn up again. The area of its mouth was 0·1 square metre. Therefore it fished through 2 cub. metres of sea, but all the water that entered the net did not again pass out through the pores of the silk. It was calculated that only 1½ cub. metres did so. Also it was estimated that the silk net only retained one-third of the diatoms that were contained in the water. That is 1½ cub. metres of water from the Bay of Kiel contained about 9000 millions of diatoms, or 6000 to each cub. centimetre. But a cub. centimetre contains about 30 drops and therefore *every drop of sea water from this part of Kiel Bay contained some* 200 *diatoms*. This was, of course, an exceptionally large haul. Hensen puts the average number of diatoms present in the West Baltic at about 457 millions per cub. metre, or about 457 per cub. centimetre. This estimate applies to the period of maximum abundance, and to the

---

[1] Except the bacteria, of course.
[2] "Beiträge zur Kenntn. chem. Zusamm. Planktons," *Wiss. Meeres. Kiel Komm.* Bd. III. Abth. Kiel, 1898.

individuals of the genus *Chaetoceros*. At the same time there were also present some 102 millions of *Rhizosolenia* per cub. metre, or 102 specimens per c.c. These figures really underestimate the richness of the sea in diatom life, for it is certain that many of these organisms living in the sea were too small to be retained by the silk of the nets used. Speaking quite generally one may estimate the number of diatoms which inhabit the North Sea or Baltic beneath every square metre of surface as anything between one and four hundred millions.

Peridinians. These protozoa are less abundant than the diatoms. We see from Apstein's results that *Ceratium*, the commonest of the peridinians, was present in the eastern side of the North Sea to the extent of about $1\frac{1}{4}$ millions in February and about $\cdot 2$ millions in November, for every square metre of sea surface. According to Hensen's results *Ceratium* is present in the West Baltic in greater abundance than in the North Sea. The average number was about $1\frac{1}{2}$ millions per cubic metre, or three individuals to every cubic centimetre. The maximum number counted was thirteen per cubic centimetre. *Ceratium tripos*, the common peridinian, is, however, only one species of many that are found in the plankton, and hosts of the smaller forms must pass through the meshes of the nets. One may estimate their abundance in the North Sea and Baltic as anything between one and ten millions per square metre of surface.

Copepods. These animals are not present in the sea in anything like the abundance of the diatoms or peridinians. But they are much larger, and their chemical composition indicates that they are of very great importance as a source of food for the larger animals. One single copepod is (on the average, for they differ greatly in size) about equal to 160 *Ceratia* or about 1500 *Chaetoceros* cells. They are relatively very uniformly distributed throughout the sea and are ubiquitous, for one finds them at all times and places. Apstein's tables shew that their abundance in the North Sea is about one-quarter to one million per square metre of surface, the numbers varying with the season. Hensen found that they were present in the West Baltic to the extent of about 80,000 per cubic metre. If we suppose that this represents the average copepod population of the sea over large areas, and the fairly regular distribution of these animals justifies this assumption to some extent, then one square mile of the Baltic contained from 80 to 100 billion of copepods. Or putting it in another way, every cubic inch of the waters of that sea contained one copepod. The estimates of the abundance of copepods furnished by quantitative plankton investigations are more reliable than those of diatoms or protozoa, for the crustaceans are more evenly distributed throughout the sea, and they are so large that all are retained by the meshes of the silk nets employed.

The micro-plankton. Thus even the comparatively large organisms, like the larger diatoms and protozoa, and the copepods, may be present in the sea in very great abundance. But there are also hosts of smaller organisms, such as the bacteria, the smaller diatoms, protozoa like the flagellates, and other minute unicellular animals and plants which are not taken in the quantitative plankton nets in the proportion in which they exist in the sea. I have already referred to the investigations of Lohmann on the density, and the means of collection, of the micro-plankton which usually escape capture in the nets made from Müllergaze. By using hardened filter-paper and thick taffeta silk as a filtering medium, and still more by utilising the contents of the filtering apparatus of appendicularians as a measure of the numbers of the smaller planktonic organisms of the sea, Lohmann shewed that there was a wealth of life in the sea not hitherto disclosed by the older plankton fishing apparatus. In the following table[1] I give some of Lohmann's results. The numbers of the organisms grouped under the term "microplankton" present in one cubic metre of the water of the Mediterranean off Syracuse are given. Those in the first column were taken by means of silk bolting cloth No. 20, and those in the second column by all other means (hardened filter-paper, taffeta silk, and by the filtering apparatus of *Oikopleura*). The numbers in the third column represent the numbers of the organisms actually present in the volume of water dealt with.

[1] The table is slightly rearranged from Lohmann's Table XIII., *Neue Untersuchungen*, &c., *loc. cit.* p. 72.

Planktonic organisms in 1 cubic metre of water in the open Mediterranean Sea off Syracuse.

	Caught by Müllergaze, No. 20	Caught by other means	Totals
1. " Producers[1] ":			
Diatoms	108,000	992,500	1,100,500
Peridinians	2,315	439,385	441,700
Flagellates[2]	80	38,600	38,680
Halosphaera	360	7,400	7,760
Other Phycozoa[3]	0	494,100	494,100
2. " Consumers[1] ":			
Rhizopoda	2,650	3,335	5,985
Flagellates[4]	20	264,380	264,400
Ciliates	690	54,435	55,125
Metazoa[5]	5,760	11,655	17,415

Total organisms in 1 cubic metre of water = 2,425,665.

The table illustrates the defect of the older plankton collecting methods. Lohmann estimates that from 28·5 % to 88·5 % of the smaller organisms escape capture by the quantitative vertical nets. These captures were made in the Mediterranean, which is a plankton-poor sea, and it is probably the case that if the same observations had been carried out in the northern seas a greater proportion of the metazoan organisms (crustacea, &c.), would have been captured. One sees, however, that the ordinary plankton apparatus, however well adapted it may be for the quantitative estimation of the larger microscopic animals and diatoms, fails to some extent when applied to the enumeration of the total plankton contents of the sea. All the tables hitherto given therefore really underestimate the abundance of life in the sea.

The larvae of the plankton. The diatoms, protozoa and copepods belong to the permanent plankton. Being adult organisms they never enter into benthic or nektic phases. But the ordinary microscopic life of the sea also includes the eggs and larvae of animals and plants which, in their adult stages, either swim

[1] See Chap. X.
[2] Flagellates with plant-like mode of nutrition.
[3] Anomalous organisms ("plant-animals").
[4] Flagellates with animal-like mode of nutrition.
[5] Includes all multicellular animals.

about in the sea, or live permanently attached to the bottom. A catch made with any kind of plankton net will usually contain a number of such eggs and larvae. For a time these drift about as constituents of the planktonic life of the sea, but at some stage in their life-history they abandon this transitory habit to take up their definitive mode of life among the benthos or nekton.

These transitory planktonic organisms are never so abundant as the three groups mentioned above. If a tow-net is used in inshore waters sometimes a haul will be made which consists almost entirely of the larvae of some invertebrate benthic animal, but such collections are not often made. One may now and then find the tow-net to contain only the zoea or megalopa larvae of crabs, or the plutei of some sea-urchin, or the veliger larvae of some mollusc. Thus Hensen records some hauls in the West Baltic which indicated that the sea in the neighbourhood of the place where the net was used contained some 170,000 larvae of the common mussel (*Mytilus edulis*) per square metre of surface. Sometimes these catches of larval planktonic animals may give us some indications of the density, on the sea bottom, of the adult creatures which gave birth to them. If, for instance, it had been known what were the exact limits of distribution of the shoal of *Mytilus* larvae to which I refer above, it would have been possible to calculate their approximate number; and if it had been known what was the average number of eggs spawned by an adult mussel, and if the probable destruction, by natural enemies, of these larvae could have been estimated, then it might have been possible to estimate the number of adult mussels on the sea bottom of the region inhabited by the larvae. The possibility of such estimations of the density of nektic and benthic animals is the chief object of the quantitative study of the larvae of the plankton.

Density of fish-eggs and larvae. Hensen's investigations of the density of fish ova are among those results of quantitative plankton work which have been most discussed in this country. Because of their economic importance the fishes have been studied more carefully than any other marine animals, and we know a great deal about the life-history of the more common species. We know, for instance, the average numbers of eggs annually spawned by the cod, plaice, haddock, whiting, and so on; the average duration of the incubation periods of the eggs belonging to these species; the spawning periods; the characters of the eggs and larvae; the ratio of the sexes in the adult phases; the rates of growth; the distribution, and so on. If it is possible to estimate the actual numbers of the eggs belonging to edible species of fishes inhabiting a sea of known area, which are produced during the entire spawning season of the year, then it is also possible to estimate the actual numbers of adult fishes of the same species inhabiting the same area—not by attempting to calculate (as Haeckel[1] supposed that Hensen did) what numbers of adults would result from the growth of these eggs, but by calculating what numbers of mature females had produced the eggs. This is what Hensen and Apstein attempted in the "Nordsee Expedition" of 1895[2].

Such an investigation presents fewer difficulties than are encountered in other quantitative plankton researches. Fish eggs are comparatively large organisms, and therefore nets of fairly large mesh (No. 3 Müllergaze) can be employed so that the filtration capacity of the apparatus can be calculated more accurately than in the case of the finer nets employed to catch diatoms and protozoa; the eggs studied are specifically lighter than water and float comparatively near to the surface, so that it is easier to work the net and obtain fair samples; and they are quite immotile and are not heliotropic, and so are more equally distributed than are copepods, diatoms or larvae. The spawning period of most sea fishes lasts about two months, but there is always a maximum during which more eggs are produced than at other times. The incubation period of most fish eggs varies from a week to a fortnight, according to the species and temperature of the sea. Suppose that a plaice egg takes exactly fourteen days to develop: at the end of this time the larva hatches out and the egg disappears. The average number of plaice eggs found per square metre of sea must then be doubled to give us the total average number produced during February. By making very frequent observations by considering the incubation periods of each species of egg captured; and by taking into account the influence of temperature on

[1] In the *Plankton-Studien*.
[2] *Wiss. Meeresunt. Kiel Komm.* Bd. II. Heft 2, 1897.

the duration of the incubation period, Hensen and Apstein were able to estimate the total number of the eggs of various species spawned in the North Sea during the spring of 1895.

The Nordsee Expedition was made during the period 15 February—1 May 1895; thus including practically the whole time of reproduction of the species of fishes investigated. These fishes were cod, haddock, flounder, plaice, dab and long rough dab. Three cruises were made: 15—22 February; 27 February—9 March; and 23 April—1 May. On each cruise a different course was set so that the vessel covered the whole North Sea with a network of lines of observation. The whole time spent in fishing was $25\frac{1}{2}$ days; and 3397 nautical miles of high sea were traversed, and 167 hauls of the quantitative plankton net were made. The fish eggs contained in each catch were identified and counted and the results were expressed as the average numbers contained in the sea per square metre of surface. Each average was calculated from the results of 167 hauls spread over 3397 nautical miles of sea, and over $2\frac{1}{2}$ months of time. Not only were the individual eggs counted but the stage of development of the embryo was noted. These stages were (1) blastoderm, (2) embryo just visible, (3) embryo marked off, (4) rudiment of eye formed, (5) eye fully formed, (6) larva hatched out. It was of importance to record the eggs found which were in the first stages of cleavage, since an unusual abundance of these would be indicative of an aggregation of spawning fishes on the sea bottom, and therefore of a "spawning place." If such an abundance of recently spawned eggs were encountered the instructions given to the naturalists were to cruise round the spot where the haul was made and determine the limits of distribution of the shoal of ova. Then the trawl was to be used and the density of spawning fishes on the bottom investigated. It is impossible to overestimate the skill with which this research was carried out, and the care which was taken to avoid error in the results.

I have given the details of the catches made in these cruises in the Appendix and the reader will see that fish eggs were taken in 158 out of 167 hauls. Now let us consider the case of the cod eggs obtained. In cruise I. 57 hauls were made and 18 of these were negative. In cruise II. 48 hauls out of 51 yielded cod eggs. In cruise III. 59 hauls were made and 13 were negative. The average numbers of cod eggs captured were:—

I. 21·4
II. 94·7   } per square metre of sea investigated.
III. 16·4

Now consider the relation of eggs in different stages of development. If we divide all the ova taken into three groups we have the following ratios:—

I. 1st stage : all later stages : larvae = 100 : 114 : 2·7.
II. do. : do. : do. = 100 : 167 : 18·3.
III. do. : do. : do. = 100 : 221 : 141.

The well-known course of a spawning season is as follows at the beginning of the season few fishes are spawning and the number of eggs per unit area of sea is therefore small; further, most of these will be in the earlier stages of development. At the middle of the period the maximum amount of spawning occurs; there is therefore a maximum number of eggs per unit area; and a greater proportion of these are in the later stages of development. At the end of the period the numbers of spawning fishes fall off; the number of eggs per unit area decreases; and there is a greater proportion still of later stages and larvae. Given a spawning period, the duration of which is much longer than the incubation period of the eggs, and this sequence must occur. But just these relations between the numbers of cod eggs taken on the three cruises, and the same relations between the various stages of development, are exhibited in the above synoptic tables; and these facts are evidence that the results of the hauls represent closely the natural conditions, which our knowledge of the life-history of the cod would lead us to expect.

Now from the results of these quantitative plankton hauls Hensen and Apstein calculated that there were produced in the North Sea, during the spawning season of 1895, 354·8 cod eggs per square metre of surface; but the area of the latter is approximately 547,623 millions of square metres, and therefore the total number of eggs produced over the entire area in the spring of 1895 was 194,297,000,000,000.

We know from actual enumerations what is the average number of eggs annually produced by a ripe female cod. Fulton[1] determined this in 1890 in the course of a laborious investigation into the average numbers of eggs annually spawned by the various edible fishes. The average number of ova produced annually by a mature female cod is 4,398,700. Now we find a certain number of cod eggs in the North Sea in the spring of 1895. If then we divide this total number of eggs produced during the breeding season of the fish by the average number known to be spawned by each ripe female, we can calculate the total number of fishes of that category present on the sea bottom during the spawning period—that is to say, *the number of fishes which have produced the ova.*

Hensen found this total number of ripe female cod in the North Sea during the spring of 1895 to be 44,172,000.

Numbers of edible fishes in the North Sea. In just the same manner, that is from a knowledge of the total numbers of ova produced during the spawning period (determined by quantitative plankton observations) and from a knowledge of the fecundity of the fishes considered (determined by actual observations), Hensen and Apstein estimated that there were present in the North Sea during the spring of 1895:—

Spawning female fishes	Numbers present
Cod	44,172,000
Haddock	180,239,000
Plaice	103,240,000
Flounder	37,807,000
Dab	772,700,000
Long rough Dab[2]	68,161,000
Total for six species	1206,319,000

Assuming, for the moment, that these figures are accurate we will proceed to expand them. They apply only to the mature spawning females of the species mentioned: the total numbers of each species are obviously much greater. To the numbers given

[1] *Ann. Rept. Scottish Fishery Bd.* ix. Pt. 3, p. 254, 1891.
[2] *Drepanopsetta limandoides.*

must be added the numbers of ripe male fishes, and the numbers of immature fishes of both sexes. These quantities can be calculated from a knowledge of the ratios of females to males, and from the ratio of mature to immature fishes of the six species dealt with. The material for the construction of these ratios exists in the fisheries literature. Fulton determined the ratios of the sexes in the case of most edible fishes, and for the six species in question, it is

$$\text{Females : males} = 280 : 100.$$

But it is much more difficult to estimate the ratio that mature fishes bear to immature ones. A good deal of work has been done in this direction but the results are not collated. In 1894 Holt[1] calculated that about 7,084,000 mature plaice were landed at Grimsby during the period April 1893—March 1894, and that 9,166,000 immature plaice were landed during the same twelve months. Therefore more immature plaice are caught than mature ones, and the same is true with regard to the other species. The ratio varies with the season, the locality, and the method of fishing; thus the whole inshore plaice fishery is practically one for immature fish, and enormous numbers of the smaller fishes are landed at some ports (London, for instance), while very large quantities are sometimes caught by the trawlers and are not brought ashore at all. Generally speaking the numbers of immature individuals of a species are greatly in excess of the mature individuals. I think that for the North Sea, and for the six fishes mentioned, we may assume that the ratio of mature to immature individuals is about 1 : 5. This estimate may not be accurate but it cannot be very far out. Considering then both the mature males, and the immature fishes of both sexes our estimate of part of the fish population of the North Sea in 1895 becomes:—

Mature female cod, haddock, plaice, flounders, dabs and long rough dabs	1206,319,000
Mature males of the same species	430,000,000
Immature males and females of the same species ...	8180,000,000
Total population...	9816,319,000

[1] Holt, *Journal Marine Biol. Ass.* Vol. IV. p. 414, 1895—7.

Here then we have a preliminary estimate of the total numbers of some kinds of fishes inhabiting a definite sea area. To what extent does it represent the actual conditions? I think that it can only be regarded as a first essay at such a census. In the first place the accuracy of the results depends on the accuracy of the estimate of the average number of eggs present in the sea per square metre of surface—that is, on the exactitude of the fishing operations involved in the capture of these eggs; and on the validity of the deduction of the average from the individual numbers— that is, on the degree of inequality of distribution of these planktonic fish eggs. Now if the reader will consult the Appendix, still better if he consults the original tables given by Hensen and Apstein, I think he will conclude that this inequality of distribution was not so great as to rob the estimated average of a certain amount of value. If we take the total number of eggs of our six species I think that the average number found per square metre may be regarded as an approximately accurate one, and that the probable error of the estimate may be calculated. If this is the case, then the pelagic estimate of the total number of fish eggs in the North Sea must be regarded as a rough approximation to the truth; and again, if this is the case, then it is justifiable to calculate from this the number of adult fishes in the same area.

But when Hensen and Apstein made their investigations the characters of the various species of pelagic fish ova were not accurately determined. It is always difficult to identify a planktonic fish egg, and in 1895 our knowledge of the life-histories of fishes was not exact enough to enable this identification to be made with certainty. Hensen and Apstein, therefore, confused eggs belonging to different species[1]—the cod and plaice for instance; and we must conclude that the numbers representing the proportions in which the six species, cod, plaice, haddock, flounder, dab and long rough dab occur, have little or no real value. But so far as concerns the *total number of all the six species*, it is probable that the results are rough approximations to the truth.

How close is the approximation? We have reasons for concluding that the formidable numbers given in the previous pages are not blind guesses at the truth, but have some degree of prob-

[1] See Hensen, *Wiss. Meeresunt. Kiel Komm.* Bd. v. Abth. Kiel, p. 157, 1901.

ability. Consider the quantities of fish landed in one year from the North Sea. In 1904 there were[1]:—

Cod...............	70,554,148	kilogrammes
Haddock	166,391,042	,,
Plaice............	58,461,073	,,
Dabs	9,551,591	,,
Flounders	1,982,415	,,

Dabs and flounders are probably much underestimated.

Long rough dabs are apparently incorrectly estimated.

Now the relation of the weights and lengths of these fishes is known, and also the average lengths of the fishes usually landed. "Equivalents" are given in the *Bulletin Statistique* which enable one to convert quantities of fish into numbers of individuals. The conversion is, in some cases, only a roughly approximate one but in other cases it can be carried out very exactly. D'Arcy Thomsson first shewed what the relation of weight and length was in the case of the plaice and Heincke and Henking[2] have given a formula for this fish which is very convenient. It is

$$\text{weight (in grams)} = \frac{\text{length}^3 \text{ (in cms.)}}{100} \times \text{constant}.$$

The constant varies from 0·8 to 1·0. It depends on the condition of the fish, which varies with the season of the year.

The numbers of fishes, of the five species mentioned, landed in 1904 from the North Sea were:—

Cod...............	65,192,000
Haddock	274,540,000
Plaice............	229,160,000
Dabs	37,279,000
Flounders	7,760,000
Total ...	613,931,000

Basing our calculation upon Hensen and Apstein's results we found that about 9000 millions of these five fishes might have

---

[1] The figures for 1895 cannot accurately be ascertained. Those for 1904 are taken from the *Bulletin Statistique* of the International Fishery Council, Vol. I. 1906.

[2] In "Schollen u. Schollen Fischerei, S. O. Nordsee," in *Beteiligung Deutschland Internat. Meeresforschungen*, IV.—v. *Jahresber. deutsch. Wiss. Comm.* Berlin, 1907.

been living in the North Sea in this year. But only 600 millions have been captured. Is this what one would expect? I think it is, for our 9000 millions might easily include one-third or one-half of that number of small unmarketable fishes which would either not be caught, or, if caught, would be returned to the sea. If it includes one-half, then the fishermen captured only about 13 per cent. of the fishes actually present.

There are no means of ascertaining exactly what proportion of the fish which frequent a fishing ground during a whole year the fishermen do catch. It is pretty certain that when a trawl-net is dragged through a shoal of fish on the sea bottom it only catches a few of those which lie in its way. Heincke and Henking (in the paper cited) marked about 600 living plaice and then liberated these on a restricted area of the North Sea, and trawled with the object of recapturing their marked fishes. But the results were inconclusive. Marking experiments of this nature have been carried out on a rather extensive scale, both in England and on the continent of Europe, and the results seem to indicate that from 10 to 25 per cent. of the marketable plaice present on a sea bottom throughout an entire year are caught by the fishermen. But so many uncontrollable factors affect experiments of this kind that their results are problematical ones.

Density of population on the sea bottom. If we assume again that our estimate of 9000 millions of cod, haddock, plaice, flounders, dabs and long rough dabs represents roughly the population (with regard to those species) of the bottom of the North Sea, then we may calculate the density of distribution. The area of the whole sea is 547,623 millions of square metres, so that if we divide 547,623 by 9000 we obtain the average area inhabited by each fish. This is about 60 square metres, or each fish inhabits a square of sea bottom which measures about 8 yards along each side. Here and there the density will be much greater, and at other places it will be much less. Those species form the bulk of the bottom-living fish population, and even when we remember that other edible and inedible fishes and invertebrates are present also on the sea bottom, the estimate of the density of life on the sea bottom does not seem to me to be an improbable one. But we must not forget that there are a considerable number of fishes, like the herring and mackerel, which inhabit the upper layers of the sea.

Very many observations point to the conclusion that the density of life, both fishes and invertebrates, is greater on the sea bottom in close proximity to the land than in deeper water further out to sea. It is in such inshore waters that one finds the greatest wealth of animal life, and I may give, as an instance, some figures relating to the density of fishes and invertebrates on a comparatively small part of the sea bottom just outside the estuary of the river Mersey in Lancashire. The results of several hundreds of hauls with a shrimp trawl made during the years 1893—1899 on this ground shew[1] that the average number of fishes of all kinds captured per haul of the net was about 5000, of shrimps 4500, and of other invertebrates about 2500. Thus the total number of animals captured per haul was about 12,000, and very often much larger catches than this were made. The length of the average haul of the trawl net was about two miles, and the width of the mouth of the net was twenty-one feet; thus in each drag an area of about 21,000 square metres was swept, and one animal was captured for every two square metres of sea bottom. Now we may be fairly sure that not more than one-fifth of the larger animals present on the bottom was actually captured by the net; and we may also be sure that a considerable fraction of the total population of the sea bottom consisted of small animals, above microscopic dimensions of course, either resident on the floor of the sea, or burrowing in the sand and mud, which were too small to be capable of capture by a net having the mesh of a shrimp trawl. I think it quite probable that there were on this part of the sea bottom not less than twenty, and not more than two hundred animals varying in size from an amphipod ($\frac{1}{4}$ inch long) to a plaice (eight to ten inches long) on every square metre of bottom.

Of course the variability of density of life on the sea bottom in inshore waters is so great that these numbers are often vastly exceeded. Thus part of a mussel bed may have a population of 16,000 molluscs to every square foot (say 1000 sq. cm.) and at times, when such a mussel bed is being devastated by star-fishes, the sea bottom may literally be carpeted by these animals. Some-

[1] *Ann. Rept. Lancashire Sea-Fish. Laby.* Liverpool, 1900, p. 39, 1901.

times the stones of a sea beach may be entirely covered with barnacles (*Balanus*), or periwinkles. But, of course, such instances of great density of population, and also those due to shoals of gregarious fishes like herrings or sprats, are rather exceptional and do not represent the average density of life of an open sea area. Just as in the case of the plankton we find that the density and composition of life depend on, and vary with, the physical conditions; so also we find that the abundance and nature of both the benthic and nektic fauna are variable, and that the amount of variation is greater the nearer we are to the coast.

CHAPTER IX.

THE PRODUCTIVITY OF THE SEA.

It should hardly be necessary to warn the reader that he must not interpret the phrase "Census of the Sea" in too literal a manner. The enumeration of the population of a country with respect to the age, sex, occupation, &c., of the individuals composing it is a process which is carried out with a high degree of accuracy, and I do not for a moment suggest that any enumeration of the individual organisms inhabiting a part of the sea can be made with such pretensions to accuracy. The materials from which such enumerations are made are (1) quantitative plankton investigations; (2) fishing experiments; and (3) the commercial fishery statistics. Now the plankton investigations claim to give only rough approximations to the truth: fishery experiments cannot be made so as to give absolute values for the numbers of fishes and other animals residing on the sea bottom; they can only afford relative values for the density of life in the sea; and the commercial fishery statistics are so imperfectly collected that their use for any scientific purpose generally leads to disappointment. What then is one to do if he wishes to form any idea as to the quantitative distribution of organisms in the sea, or in fresh water? With such imperfect methods and data only rough approximations can be made. It is no service to science only to urge "counsels of perfection," one should rather make use of what data are available, and trust that the provisional results thus attained may assist in the further elaboration of methods of investigation; and even if we are unable to give accurate figures for the population and productivity of the sea, it is always of interest to know what are the approximate values. We may not be able to find out, Hensen says, whether our neighbour's income is £1000 or £1200 per year, but we are interested in knowing whether he is a millionaire, is wealthy, or is only in comfortable circumstances.

Our estimates of the abundance in the sea of diatoms, protozoa, copepods, fishes and so on, are of just this degree of value. But even when we have obtained these approximate figures for the population of the sea it is not enough, for such populations are continually changing. Organisms die and fall to the sea bottom and decompose, or are devoured by their enemies or are captured by man. Birth-rates in the sea vary with each kind of organism and change with the season, and the rates of growth undergo corresponding fluctuations. Death-rates too change with the season, and with changes in the density of inimical organisms. Not only must we attempt to estimate the density of population in the sea at a given time, but we must also try to find out what mass of living substance is periodically generated.

It is much more difficult to attempt such estimations of the productivity of a sea area, than merely to attempt to ascertain the mass of life at one particular time. Man has hardly at all cultivated the sea in the way that he cultivates the land. When compared with the present position of agriculture there is hardly any science of aquiculture. With the exception of certain attempts by Hensen and Brandt to estimate the productivity of certain inshore sea areas in the Baltic, and a very incomplete study, by the Germans, of the conditions of culture in carp-ponds, the science of aquiculture is practically non-existent. It is true that both the French and Dutch cultivate oysters on quite a large scale, and the Americans deal very largely with the artificial fertilisation and rearing of marine and fresh-water fishes, and in this country trout and salmon hatching is carried on to some extent. But the French and Dutch oyster farms are conducted in apparently a purely empirical manner, and both American and English fish culture is as yet quite experimental in its aims. Agriculturists have acquired much information as to the conditions of cultivation of crops, the metabolic processes involved in the rearing and fattening of live stock, and so on; and it is known what mass of produce in the form of cereal and other crops and live stock can be "raised" from unit areas of different soils under different conditions, but such knowledge hardly exists with regard to marine economic produce.

I will collect here what data are available with regard to the productivity of the sea. One turns naturally to the fishery statistics in attempting such estimations. If we can ascertain what quantity of fish is landed annually from the North Sea, we may have the data for a first essay at the productivity of a fishing area. A year or two ago such information was not very accessible but during the last two years the International Council for Fishery Investigations have published fairly complete statistics and I quote these here[1].

Productivity of the North Sea fishing grounds. In the year 1904 the fishing fleets of Russia, Norway, Sweden, Denmark, Germany, Belgium, the Netherlands, England, Scotland and Ireland caught in the North Sea and landed at European ports 967,433,000 kilogrammes, or 951,900 tons of fish. The area of the North Sea is equal to 547,623 millions of square metres[2]; we find then that the yield of this fishing area was (in 1904)

17·6 kilogrammes per hectare per annum,

or, putting the same thing in English measures:

15 lbs. per acre per annum.

If we take the average value of the fish caught as about 0·234 shilling per kilogramme, then the money value of the North Sea was 1s. 3d. per acre per annum.

Now this is certainly a minimum value for the yield per unit area of the North Sea in fish flesh. Fishing boats do not go to sea to make scientific deductions but to earn money for their owners, and they only bring to land such products as command a ready sale in the markets. Thus hosts of small and inedible fish are caught and immediately shovelled overboard. Shellfish (lobsters, crabs, shrimps, molluscs, &c.) are not included, for the difficulties in the way of estimating the catch of these creatures are too great. Probably the quantity of fish included in the statistical returns is

[1] See *Bulletin Statistique des Pêches Maritimes des Pays du Nord de l'Europe*. Vol. I. Copenhague, 1906.
[2] Karsten, "Neue Berechnung mittl. Tiefe Ozeane," *Dissertation*, Kiel, 1894.

underestimated. The reader must therefore regard this yield of 17·6 kilogrammes per hectare as only something to start out from in our investigation. When compared with other reliable estimates of the productivity of a sea area it appears very small.

The North-Western fishing grounds. The North Sea is only one fishing area, and it may interest the reader if I give here the total (minimally estimated) value of the fisheries of the North European nations[1]. The figures for the year 1904 are :—

Quantity and value of fish landed from the fishing grounds of North Europe.

	Quantity (kilogrammes)	Value, £
Russia	4,488,460	92,229
Sweden	990,892	317,937
Norway	311,614,192	1,629,389
Denmark	37,434,824	571,421
Germany	61,479,838	664,698
Netherlands	123,219,867	997,118
Belgium	2,899,590	86,106
England	577,346,267	6,779,987
Scotland	403,749,711	2,307,902
Ireland	48,353,268	393,630
Total	1571,576,909	£13,840,417

That is, about 1½ millions of tons of edible fishes worth about 14 millions of pounds were taken in the year 1904 from the fishing grounds of Northern Europe.

Productivity of an inshore fishing area. Generally speaking one would expect a greater return from the fisheries of a part of the sea lying near the land than from an open sea region. This is because the greater bulk of the fish taken from the North Sea are caught by means of trawl nets, lines and drift nets, while in an inshore fishing area not only are these methods employed but there is also a great amount of fishing by smaller boats for shrimps and prawns, and also a considerable amount of stake

[1] See *Bull. Stat.* The quantities do not include shellfish landed. The money values do.

net fishing. Since there is also a longer coast line relatively to the whole area, the foreshore fisheries for shellfish and by means of stake nets are also relatively greater. Probably, too, the fishing is generally more intense in an inshore, than in an open sea area. We may take Morecambe Bay on the west coast of England as a good example of such an inshore fishing area. There we have trawling, some lining, and a very considerable fishery for shellfish. The area of the Bay, including the foreshore, is about 155 square nautical miles (or 53,000 hectares, or 131,000 acres). In the year 1906 there were caught in the Bay, and landed at the fishing ports on its shores, 4680 tons of all kind of fish[1]. Dividing this by the total area we find that the yield was

     89 kilogrammes per hectare per annum;
or   79 lbs. per acre per annum;
and the value is about 2s. 6d. per acre per annum.

Productivity of some inshore shell-fisheries. We see then that an inshore fishing area is usually more productive than an open sea one, partly because there is, as a rule, more fishing carried on near the shore than well out at sea, and also because there is, as a rule, a more abundant fauna in the shallow waters near the shore. Now we may narrow down our enquiry still further and try to ascertain what is the yield of a shellfish area such as a shore containing many mussel or cockle-beds.

Reliable figures are available for the mussel fisheries at Conway, on the coast of North Wales, and Morecambe, in Morecambe Bay in Lancashire. If we take the case of the Morecambe mussel fishery we find that the area covered with mussel beds is about 578 acres, or 234 hectares in extent. In 1906 the quantity of mussels taken from these grounds was 49,908 cwts., or 2,540,000 kilogrammes[2].

[1] The year 1906 is the first one for which accurate statistics are available. Prior to this date statistics were collected by the Board of Trade and the Board of Agriculture and Fisheries but the values published are probably low. In 1906 the Lancashire and Western Sea Fisheries Committee began to supervise the collection of data and reliable returns were obtained. I am indebted to Dr Jenkins, of the Lancashire Committee, for the figures.

[2] *Lancashire and Western Sea Fisheries, Superintendent's Report*, 31 December 1906.

CH. IX] THE PRODUCTIVITY OF THE SEA 183

Therefore we have a yield of
> 10,860 kilogrammes per hectare per annum;

or 86 cwts. per acre per annum;

and the money value is £14. 16s. per acre per year.

In the case of the Conway mussel fishery the area exploited is about 256 acres, or 103·6 hectares. In 1906 the total quantity landed was 22,252 cwts., or 1,130,000 kilogrammes[1]. This gives us a yield of
> 10,900 kilogrammes per hectare per annum;

or 87 cwts. per acre per annum;

and the money value is £14. 18s. per acre[2].

In 1899 I estimated[3] the total produce of the Lancashire cockle fisheries as about 6685 tons for the year. This is probably well above the average and the product varies greatly from year to year. The total extent of cockle-bearing sands I placed at about 105 square nautical miles, or 35,900 hectares. All this extent of sand does not always produce cockles. There is, in fact, a kind of rotation in the fishery: at one time certain areas are densely populated with the molluscs and the fishery is then very intense on these restricted portions of the whole area. Soon, however, these beds become exhausted of the larger, legal-size cockles and the fishery ceases for a year or more until a fresh crop of marketable shellfish grow up to maturity. The fishermen, in the meantime, seek new grounds. So in the course of time the whole area is exploited, though only a fraction of it is worked at any one time. Nevertheless the whole extent of cockle-bearing sands must be considered. But in considering our estimate it has to be remembered that only cockles of a certain size are allowed to be taken, and that if the legal size limit were greatly reduced the yield would rise. The yield was in 1889
> 189 kilogrammes per hectare per annum;

or 166 lbs. per acre per annum.

The value is about 19s. per acre.

Now all these estimates of productivity are those of uncultivated sea areas. If we consider a cultivated fishery the yield will be

[1] *Lancashire and Western Sea Fisheries, Superintendent's Report*, 31 December 1906.
[2] Conway mussels command the better price.
[3] In *Ann. Rept. Lancashire Sea Fisheries Laby.*, for 1899.

much greater. There are, however, few cultivated fisheries which can supply us with this information. Here and there in Great Britain oysters are imported from America and elsewhere and laid down to be fattened, and in France and Holland the "succulent bivalves" are actually bred and reared. But I do not know of any published figures for the yield of these oyster layings or farms. We have, however, figures for the yield of the German carp-ponds.

German carp fisheries. In Germany the cultivation of the carp is carried on extensively. In many places large ponds are stocked with the fish and the fishery is worked as a commercial affair, while in most small villages smaller fish ponds are to be found. The ponds are regularly stocked with young fish and the latter are fattened. Sewage is deliberately led into these ponds, and the resulting addition of nitrogenous inorganic food stuff sets up an abundant microscopic fauna and flora, which is the ultimate food of the fishes. Whether a carp-pond is situated badly or well, the addition of sewage always makes it better. There is quite a literature relating to the practical culture of carp, and the details of carp metabolism have been studied in much the same kind of way as in the breeding and raising of live stock on the farm. Many estimations of the yield of these carp-ponds have been made. I quote a few. Hensen[1] estimates the yield of certain ponds as 76·5 kilogrammes of carp flesh per hectare; Brandt[2] estimates the least yield of some ponds at Stettin as 65·5, the average as 106·5, and the maximum yield as 164 kilogrammes per hectare. Von Stemann[3] gives the yield of some carp-ponds in Schleswig-Holstein as 112·2 kilogrammes per hectare and year. Thus we have a productivity for a fresh-water cultivated area of 65·5 to 164 kilogrammes per hectare, or 58 to 141 lbs. per acre; and a value[4] varying from £6. 11s. to £16. 9s. per hectare and year.

Productivity of a cultivated mussel fishery. Since the yield of an inshore fishing area is greater than that of an open sea one we should also expect to find that a cultivated shell fishery

[1] "Resultate stat. Beobacht. Fisch. deutschen Küsten," 3 *Ber. Komm. Untersuch. deutschen Meeres*, Kiel, 1878.
[2] "u. d. Stettiner Haff," *Wiss. Meeresunt. Kiel Komm.* Bd. I. Heft 2, 1896.
[3] *Allgemeine Fischerei-Zeitung*, Berlin, 1895.
[4] One pound of carp is worth about 1s. in Germany.

would yield results still greater. I can only quote the results of one such fishery, but these are very instructive. At Morecambe on the coast of Lancashire there is a large area of foreshore on which "over-production" takes place; that is to say, enormous numbers of mussels are spawned, so that there is not room for all the molluscs to grow and great numbers remain permanently stunted and dwarfed, reproducing nevertheless so that the over-crowding of the mussel-beds persists. Formerly great masses of these small mussels were carted away by farmers and applied to the land as manure. Then the Fishery Committee (in some ignorance, apparently, of the problem with which they had to deal) stepped in and prevented the "removal from the fishery" of mussels which were under a certain size, which they fixed by regulation. So the depletion of the beds ceased and as a consequence they came to contain enormous numbers of mussels which, being stunted and under the legal size, had absolutely no economic value. The Fishery Authority now proceeded to apply the logical complement to its restrictive legislation and encouraged the fishermen to "transplant" the undersized shellfish. Great numbers of the molluscs were therefore removed from the overcrowded beds and re-deposited in a locality—"Ringhole," where the conditions were known to be such as to favour the growth of the shellfish[1].

Now the great theoretical interest of this experiment was overlooked so that there is, unfortunately, not so much data regarding it as one would wish for. I attempt an estimate here which is probably a fairly approximate one. I may observe that the experiment was a very decided success in a commercial sense and was warmly welcomed by the fishermen, who derived great benefit from its results. In April 1905, some 347 tons of dwarfed mussels were taken from the overcrowded beds and put down in Ringhole. The fishing of this area was then prohibited by mutual agreement among the fishermen until November of the same year, when the transplanted shellfish were taken for the markets.

[1] See Scott and Baxter, *Ann. Rept. Lancashire Sea Fisheries Laby.*, Liverpool, for 1905, for an interesting account of these operations. See also *Superintendent's Reports Lancashire and Western Sea Fisheries Committee*, 1906–7, Preston.

The stunted mussels measured, when transplanted, $1\tfrac{5}{8}$ to $1\tfrac{7}{8}$ inches in length and 4634 bags were re-deposited. Each bag contained on the average some 7000 mussels. Therefore about 32 millions of the shellfish were dealt with.

Now we cannot assume that all these mussels survived in their new home. Probably not less than half of them did so. The rest were doubtless smothered by sand and mud, or destroyed by starfishes, boring molluscs, sponges, worms, &c. We will assume that 16 millions survived and underwent growth.

The average weight of the dry organic substance (excluding that of the shell) of one of the stunted mussels $1\tfrac{7}{8}$ inches in length was 0·598 gram. The average weight of dry organic substance in one of the same mussels of $2\tfrac{1}{2}$ inches in length *after* transplantation and growth (also excluding the shell) was 1·311 gram. Therefore the average gain in dry organic substance was 0·713 gram per mussel[1]. This represents the growth during a period of about eight months. Probably this period (April—November) represents the principal growing period of the year, but nevertheless there was probably some growth also during the winter months. The area of Ringhole is 10 hectares or 25 acres. Now calculating the gain on 16 millions of mussels we find that the productivity of Ringhole *in dry organic substance* was

1140 kilogrammes per hectare;

or 9 cwts. per acre.

I do not doubt that this is a minimum estimate. We must remember that there is a certain amount of fishing for sea-fish in the same area. This is small but only because the fish which might have been caught there (dogfishes chiefly) had little economic value. Then a very considerable bulk of invertebrate life, and also planktonic life, not capable of utilisation by the mussels, must also have been produced. There is a strong tidal flow in and out of Ringhole, and since more of the ultimate food stuff of animals (plankton) was probably produced in Ringhole than in the open sea of Morecambe Bay it is the case that the inshore area was probably depleted to some extent by the tidal circulation.

---

[1] This weight will obviously vary with the conditions of the shellfish, i.e. with regard to spawning.

Altogether I do not doubt that the productivity of this area, in terms of dry organic substance, cannot be much less than 1500 kilogrammes per hectare.

The absolute productivity of a sea area. What then is the probable quantity of organic substance which can be produced in an inshore sea area? All the estimates quoted refer only to the quantity of directly utilisable produce which may be taken from the sea—a very different thing indeed. We wish to know what total mass of life is generated throughout the year per unit area of a sea such as the Baltic or Irish Sea. Now the estimation of this quantity is a problem of such very great practical difficulty that one may well despair of its solution. But if we assume that it is possible, by means of quantitative plankton investigations, to deduce the quantity of life present in a sea area at a definite time, then the calculation of the absolute productivity is theoretically possible. At the beginning of the year a certain mass of life is present, and at the end of the year much the same quantity is still present. But in the meantime all organisms have been reproducing and growing. The mass of life present at the beginning of the year is the capital; the mass generated during the year by the reproduction and growth of the capital is the interest. At the end of the year the capital remains the same: the interest has been eaten up, or otherwise destroyed. What is the rate of interest?

In order to determine what is the interest we require to know (1) the rate of reproduction of each species of organism under different conditions (temperature, weather, &c.); (2) the rate of growth of the individuals of each species (also under different conditions); (3) the average duration of life of the individuals of each species; (4) the duration of reproductive activity in the life-history of each species; and (5) the amount of natural destruction due to enemies. Some of these things we do know: for instance, we know what numbers of eggs are spawned by most fishes, and some invertebrates; the rate of birth in some invertebrates (thus from the relation between eggs, larvae and adults in the plankton, Hensen ascertained that the birth-rate of copepods (all) was about 134 per 1000); the duration of life in many marine organisms (thus Hensen determined that a copepod lived, on the average, for about a week, and so the rate of birth was about 134 per 1000 per week). Reasoning in this way, and considering the average numbers of copepods present per square metre of Baltic water, Hensen came to the conclusion that there were produced, per annum, 8,866,000 copepods in every 10 cubic metres of the Baltic. Since in a stationary population the birth-rate and death-rate are the same, it follows that in every 10 cubic metres of the Baltic, 8,866,000 copepods must have been destroyed annually, that is, eaten by other animals.

Now theoretically sound as this method is, it is nevertheless incredibly difficult to make the calculation of productivity; for we do not know precisely the birth-rates, rates of growth, &c. The accuracy of Hensen's calculation—that 8,866,000 copepods are annually produced per 10 cubic metres of Baltic—is therefore very problematical. But we may check this result roughly, so as to see that it is something more than a mere guess. We know that herrings feed largely (or entirely) upon copepods; we know how many herrings are caught in the course of a fishing season in a roughly determined area of sea; we know quite accurately how many copepods are contained in the stomachs and intestines of a sample number of such herrings examined; and we may estimate roughly, or assume, how often a herring uses up its own weight of organic substance. Thus we may calculate roughly how many copepods have been eaten by the herrings per unit of time and sea. I leave the sceptical reader to make these calculations and recommend him to do so before he concludes that Hensen's estimate is all nonsense.

Thus by the application of such principles Hensen determined that so much plankton was annually generated in the Baltic per year, under each square metre, as was equal to 150 grams of dry organic substance. He tells us that the value of his estimation lies more in the study of the method than in the result[1]. There can be little doubt that the result obtained by him is a minimum one. I have shewn by quite a different method, that the productivity of Ringhole, at Morecambe, in dry mussel flesh

[1] See "Fruchtbarkeit des Wassers": in "Nordsee-Expedition, 1895"; *Wiss. Meeresunt. Kiel Komm.* Bd. II. Heft 2, 1897.

was, at the least, 114 grams per square metre per year and the approximation of these results is interesting.

The composition of marine products. I have used the terms "dry organic substance," "mussel flesh," "fish flesh," &c. Now in any consideration of the productivity of a sea area, it is necessary to compare the quantities of organic substances with each other, and with corresponding land organic products. That is, we wish to know what is the composition of such marine "crops" as plankton of different kinds, fish, and other marine flesh, &c. Brandt[1] in 1898 made a number of such determinations of the chemical composition of the plankton and I quote some of his more important results here. Catches were made with a quantitative plankton net, so that it was possible to estimate the numbers of organisms present per unit volume of sea water. These catches were then subjected to quantitative chemical analysis. Some catches consisted predominantly of diatoms, others of peridinians, and others again of copepods. Thus it was possible to obtain values for the chemical composition of the principal planktonic constituents. Mixed plankton catches, characteristic of different seasons, were also analysed, so that the composition of the "meadows of the sea" could be compared with that of land crops. All Brandt's results were obtained from the examination of the plankton of Kiel Bay.

The following table gives the percentage composition of the dry substance of plankton and of various planktonic organisms:—

	Proteid	Chitin	Fat	Carbohydrate	Ash
Autumn Plankton	20·2 to 21·8	—	2·1 to 3·2	60 to 68·9	—
Peridinians.........	13	—	1·3	80·5	5·2
Diatoms	10	—	2·8	22	65·2
Copepods............	59	4·7	7·0	20	9·3

For comparison with these figures I quote the composition of various land crops[2].

[1] Brandt, "Beitr. chemisch. Zusammensetzung Planktons," *Wiss. Meeresunt. Kiel. Komm.* Bd. III. Abth. Kiel, 1898.
[2] *Ibid.*

Percentage composition of the dry substance of land crops.

	Proteid	Fat	Carbohydrate	Ash
Ordinary Meadow Hay..	8·7	1·7	83·6	5·8
Good ,, ,, ...	13·6	3·2	75	8·2
Rye (Grain)	12·8	2·3	82·3	2·1
Peas	26·4	2·2	68·2	3·1
Potatoes	8·4	0·8	87·2	3·6

Brandt also gives the composition of the bodies of various fishes and other marine animals; the exoskeletons (shells and carapaces) of the molluscs and crustaceans are not included.

Percentage composition of the dry substance of some marine animals.

	Proteid	Fat	Carbohydrate	Ash
Herring	56·42	35·85	—	7·02
Salmon	60·49	35·62	—	3·89
Flounder	87·61	4·38	—	8·0
Cod	91·08	1·86	—	7·6
Lobster	79·80	10·13	0·16	9·41
Crab	78·87	7·69	3·75	9·6
Oyster	46·8	9·5	28·1	16
Mussel	54·86	7·07	26·0	12

For comparison with this table I quote the following analyses[1].

Percentage composition of the dry substance of the whole bodies of farm animals (contents of intestines, etc. excluded).

	Nitrogenous substances[2]	Fat	Ash
Half-fat ox	44·9	43·8	11·2
Fat ox	41·6	47·4	9·6
Store sheep	40·5	51·1	8·4
Fat sheep	24	70·5	5·5
Store pig	34·5	58·6	6·7
Fat pig	20	77	3

[1] From results of the Rothamsted experiments.
[2] Include proteids, gelatinous substances and horny substances.

These analyses show that the plankton is to be compared with the cereal and other crops: it represents the vegetation of the sea. Marine animals such as fishes and shellfish are obviously to be compared with the animals bred for human food. The dry substance of the bodies of fishes contain more nitrogenous substance and less fat than one finds in the bodies of farm animals. Salmon and herring, as everyone knows, contain a greater percentage of fat than does the flesh of most other edible fishes, such as the cod and flounder. Crustacea, like the crab and lobster, contain a large percentage of proteid and little fat, but a small proportion of carbohydrate. On the other hand molluscs, like the oyster and mussel, contain much about the same proportions of proteid as does the flesh of lean farm animals, comparatively little fat, and a considerable percentage of carbohydrate.

The plankton is difficult to compare, as regards its chemical composition, with land produce. It usually consists predominantly of organisms like the diatoms and peridinians, and of copepods, or larvae which resemble these in their composition. It contains the "proximate principles," proteid, fat and carbohydrate, but the relative proportions of these vary with the kind of organisms which make up the mass of the plankton. A planktonic catch usually contains a considerable percentage of ash, which comes from the skeletons of the organisms—the siliceous shells of diatoms and other limy or flinty skeletons; and in catches which consist chiefly of diatoms, this siliceous ash is always considerable. A plankton catch consisting chiefly of copepods does not correspond in its chemical composition to that of any other animal. These little crustacea contain a relatively large percentage of proteid, and a moderate amount of fat and carbohydrate. They resemble more the composition of the oyster than that of any other common animal. If they could be obtained in quantity they might afford a highly desirable and nutritious food substance[1].

The diatoms, however, form but an indifferent "crop." The proportion of proteid is small, and that of ash very high. This is

---

[1] They have indeed formed part of the menu of a yachting party. See Herdman in *Nature*, July 23, 1891, p. 273; also *Trans. Liverpool Biological Society*, Vol. vi. p. 78, 1892, for an account of such a culinary experiment, and for some interesting suggestions as to the uses of copepoda as human food.

because the siliceous skeleton forms a relatively large part of the mass of the organism. Peridinians, such as *Ceratium* or *Peridinium*, are much more favourable, and a catch consisting chiefly of these protozoa is not unlike, in its chemical composition, that of a rye crop; and Brandt places them in an intermediate position between rye and good meadow hay.

It is interesting to compare these three predominant planktonic organisms with regard to the mass of dry organic substance afforded by corresponding numbers of individuals. Thus:

1 gram dry copepod substance corresponds to 675 millions of the diatom *Chaetoceros*; and to 42 to 65 millions of peridinians; and to 300,000 to 500,000 copepods;

or, putting it in another way:—

1 copepod contains as much dry substance as 135 peridinians, or 1687 diatoms; and one peridinian contains as much dry matter as 12 diatoms.

The relative productivity of the sea and land. We see then that the land and sea afford organic products which may be compared with others in respect of their chemical composition and suitability as human food. Fish flesh may be compared directly with such land produce as beef, mutton and pork. The marine shellfish products, such as molluscan and crustacean flesh, differ notably from the flesh of farm animals but obviously belong to the same category of produce. The plankton is comparable with the land crops, so that a mixed catch of the former does not differ greatly in composition from that of a crop of meadow hay, or from ordinary pasture. In every way, the plankton is to be compared with pasture; not only in composition but in function, since the former is indirectly or directly utilised as food by marine animals which in their turn afford a flesh product to the fisherman; while pasture is, of course, utilised in just the same way by the animals raised by the farmers. Land and sea then yield corresponding products. Is it possible to compare the relative productivities of the two regions?

There are unfortunately but slender data available for such a comparison. Agriculture is a much more highly developed science than agriculture; nevertheless there is some difficulty in estimating the absolute quantity of organic matter, in the form of crops, or flesh produced from cultivated land per unit area. One turns naturally to the Annual Returns published by the Board of Agriculture and Fisheries for this information, but only to be disappointed. We learn, for instance, that the average crop of the period 1896—1906, in Great Britain was 31·22 bushels of wheat per acre, 38·92 bushels of oats, 28·59 bushels of beans, 5·78 tons of potatoes and so on[1]; but beyond these unilluminating figures the official returns disclose little of general interest. The quantities of the various crops thus accounted for do not represent the productivity of the land, for the straw of grain crops must also be considered as well as the "crop residues" of roots, etc., and the weeds also produced per unit area. Considering both grain and straw we find (from other sources of information) that the average yield in dry substance of British crops may be put as 4,183 lbs. of wheat per acre, 3,827 lbs. of barley, 3,987 lbs. of oats, 2,822 lbs. of meadow hay, 3,461 lbs. of beans, and so on[2]. But information as to the absolute quantities of proteid, fat, carbohydrate and ash, yielded by cultivated land under different conditions is apparently difficult to obtain. There is, however, a German estimate by Biebahn and Rodewald, which puts the productivity of cultivated land as equal to 1,790 kilogrammes of dry organic substance per hectare.

The available data for our comparisons are:—

(1) *In terms of flesh:*

Cultivated land,
83·5 kgs. of beef per hectare (Biebahn);
Cultivated fresh-water carp-ponds (Germany),
65·5 to 164 kgs. carp per hectare;
Cultivated mussel beds (Morecambe),
8,000 kgs. mussel flesh per hectare;
Uncultivated cockle beds (Lancashire),
66 kgs. cockle flesh per hectare[3];

[1] "Agricultural returns," Bd. of Agric. and Fisheries.
[2] Warrington, *Chemistry of the Farm*, London, 1891.
[3] "Flesh" is about 35 % of the total weight.

Uncultivated mussel beds (Conway)[1],
4,033 kgs. mussel flesh per hectare;
Uncultivated inshore general sea fisheries (Morecambe Bay),
89 kgs. fish and shellfish per hectare;
Uncultivated North Sea fishing grounds,
17·6 kgs. fish per hectare.

(2) *In terms of dry organic substance:*
Cultivated land (Biebahn and Rodewald),
1,790 kgs. per hectare;
Cultivated mussel beds (Morecambe),
1,140 kgs. per hectare;
Uncultivated sea (Baltic), yield in plankton (Hensen),
1,500 kgs. per hectare.

From these figures it appears that the produce of a large uncultivated water area is less than that of a cultivated land area, whether we take the yield in fish or shellfish flesh, or the yield in dry substance as the basis of comparison. But a cultivated water area, such as a fresh-water carp-pond, or a part of the sea near the shore treated so as to produce shellfish under the most favourable conditions, is capable of affording a rich " crop "; and if aquiculture were as intelligently studied and practised as agriculture there can be little doubt that the sea would be more productive than the land. Thus Hensen estimated that the mass of plankton (ultimate organic substance) produced in the uncultivated Baltic was not far short of that produced upon cultivated land.

Let us attempt to compare the nature and density of life on a land surface in the temperate zone, with a sea surface of corresponding extent. If we were to explore a large tract of cultivated and forest land, in which crop-lands, meadows, woodland, streams, and moorland occurred, we should find that everywhere vegetable life would be predominant, and that animal life would be comparatively sparsely distributed. Overhead in our meadow land and in the cultivated parts would be a few birds and insects, while in the soil and on the vegetation there would occur insects, worms, and here and there rodents like mice and rabbits. In the wood and

---

[1] Taking the weight of flesh in a mussel as 37 % of the total weight.

animal life would be more abundant; insects would be present everywhere, and birds would be more numerous than in the open country, yet birds would not be resident in every tree. Small mammals, though more numerous than in the meadow land, would not be very evident. Perhaps animal life would be most abundant in the streams and lakes, but even here fishes, water insects and the aquatic mammals would not be very abundant. But everywhere vegetation would be comparatively luxuriant.

Suppose now that the waters of the North Sea were suddenly to disappear, and that the whole mass of life contained in them were suddenly to be precipitated to the sea bottom. What kind of picture would then be presented? We should see a vast, almost level plain literally carpeted with animal life. Everywhere there would be a glittering mass of fish scales, for we should see not only the fishes which live normally at the sea bottom but also those which lived pelagically, like the herring and mackerel. Hordes of invertebrates, crabs, starfishes, molluscs, &c., would be mingled with the fishes. The mud and sand would also yield their quota of living things, and these would be much more numerous than the few worms and insects contained in the land soil. Every square inch of the bottom would be heaped up with animal life, and the whole would be partially smothered by the plankton precipitated from the water. Vegetation, as it appears on the land, would be very scarce, for the sea-weeds would be confined to the coastal margin; and we should hardly recognise the plankton as of vegetable nature. The irresistible impression would be that the sea was very much richer in life than the land[1].

Such an impression would be an accurate one. For production of organic substance upon the land is restricted to the surface of the soil, and to a very thin layer of the latter; while in a shallow sea, like the North Sea, production by plants is carried on throughout a stratum of water, the average thickness of which is not less than 200 feet. Though sea-weeds only exist along the shore, and at the sea bottom near the latter, yet the vegetable plankton exists practically everywhere, and at every level of the water filling up the North Sea basin. Then we find vast tracts of dry land which are

[1] This comparison is made by Hensen in "Das Bild der Nordsee" in "Nordsee Expedition, 1895."

either utterly sterile, or are productive to a very slight degree. Such are desert-lands; the higher rocky parts of mountainous country; the enormous tracts of land covered transitorily or permanently by snow and ice; and the relatively unproductive moorlands. And even of the productive land surface only a very small part is under cultivation. But everywhere in the sea, even under the ice and in hot and cold areas, we find abundant life. No part is sterile, and the variations in productivity are, when compared with those on the land, of little account. If we take equal average areas of land and sea, we will find that the yield of the latter is greater than that of the former. Even the comparatively poor yield in fish per acre of the North Sea is probably greater than the yield per acre of *all* the land in Great Britain and Ireland.

The impoverishment of the sea. We may regard that part of the sea which is exploited by the fishing fleets of North Europe as an uncultivated water area which bears annually a certain crop of fish flesh. There are only very few spots along the shore where the culture of shellfish is attempted and the product of these cultivated sea-water areas is so small (relatively to the total yield) that it may be neglected. At least 952,000 tons of fish are taken annually from the North Sea, and about a million and a half tons from the whole north-western fishing area. This is the annual crop which the fishermen of North Europe take from the sea. It is a "harvest of the sea" which does not correspond to any sowing.

Now the necessity for all fishery legislation is based on two considerations (1) that among a number of fishing vessels and men working on a restricted fishing area, there must generally be the possibilities of disorder, that is, fishermen belonging perhaps to various nationalities, and perhaps practising different methods of fishing, must at times interfere with each other, producing disputes which may be productive of trouble, and not easy of adjustment; and (2) that since the quantity of useful fish products that may be taken from the sea is limited, the time must come when the produce of the fisheries will suffer impoverishment merely because more fish are being captured than can be replaced by the natural powers of reproduction and growth of the species sought for by the fishermen. So two categories of fishery legislation have gradually come into existence, one group of regulations designed to prevent the possibility of dispute between different classes of fishermen, and the other group intended to prevent the serious depletion of the fishing grounds. With the former we have nothing to do here, and we are only concerned with the assumption on which the second is based.

There have always been two opinions with regard to the possibility of depleting a natural fishing ground. On the one hand it is argued that the quantity of any one kind of fish in the sea is so great, the individuals are so prolific, and the amount of destruction by the natural enemies of the fish, and by physical agencies, is so enormous, that anything that man can do, in the way of further destruction by fishing operations, has no appreciable effect on the numbers of fish on a fishing ground[1]. On the other hand it is argued that man's influence is much more powerful, and both the fishery statistics, and the practical experience of those engaged in the trade, agree in the conclusion that many kinds of fish are now less abundant in the sea than was formerly the case.

Now one must remember that fishermen go to sea in order to earn money. During the last half of the nineteenth century there gradually arose an increased demand for fish food, as the means of transport developed and as great centres of population were placed in easy communication with the fishing ports. So a gradually increasing exploitation of the older fishing grounds, such as the North and Irish Seas, was developed under the stimulus of this increasing demand for the produce of these areas. Every year saw an increase in the number of fishing vessels, and then towards the 'eighties and 'nineties great numbers of powerful steam-trawling, drifting, and lining vessels were built and equipped for work in these waters. But it was soon found that the catches of fish made in the North Sea between 1850 and 1880 were no longer possible, and the continued exploitation of the

[1] This was the position taken up about 1860–70 by Huxley, Spencer Walpole and others. More recently it has been adopted by MacIntosh. See *Resources of the Sea*, Cambridge University Press, 1899.

North and Irish Seas ceased to be as profitable as in the middle decades of the century. If all that man could do had been inoperative in reducing the fish population of the North Sea, then this fishing area would have been sufficient. But it was discovered that profitable "voyages" could no longer be made there, and so year by year, since the beginning of the 'nineties, the steam trawlers have forsaken the North and Irish Seas for more lucrative fishing grounds in the Bay of Biscay, off the coasts of Ireland, off Iceland, and lately in the White Sea.

This is the "practical" argument in favour of the depletion of the North Sea fishing grounds. A complete and rigid demonstration of the decreased abundance of fish on this ground has apparently still to be made. If we had an absolutely exhaustive system of collection of fishery statistics it would be an easy matter to shew what is the condition of the North Sea relative to its condition fifty years ago. But so imperfect are our fishery statistics that no attempt at such a demonstration of depletion has yet been made that has not been unfavourably criticised. In 1900 Garstang[1] shewed (1) that the total quantity of certain classes of fish landed from this area had not increased very greatly during the decade 1889–98; (2) that the catching power of the vessels employed in fishing there had greatly increased; and (3) that, as a consequence, the average catch made by each vessel had gradually decreased. Fulton[2] had previously adopted the same line of argument with reference to the Scottish trawl fisheries. But it has been contended[3] that this method of shewing the depreciation of the fishing grounds is a fallacious one, though I cannot see in what manner the method fails.

Many attempts, both in Great Britain and on the Continent, have been made to shew that the fishermen catch a very notable percentage of the fish present on the grounds. Numbers of living plaice have been marked by means of various kinds of labels, and these fishes have been liberated in the sea. After a year has elapsed a certain percentage of these marked plaice are always recaptured by the fishermen, and it is claimed that the

[1] *Journ. Mar. Biol. Association*, vol. VI. 1900–3.
[2] *Ann. Rept. Scottish Fishery Board for* 1891, Pt. 3, p. 171.
[3] *Ann. Rept. Inspectors of Fisheries for England and Wales for* 1900, p. 5.

percentage of the marked plaice recaptured represents roughly the percentage of *all* the plaice (of the same range of size as the marked fish) caught on the same fishing grounds. That is, it has been argued, that if (as has often happened) 25 per cent. of the marked fish liberated are recaught on a certain fishing ground, 25 per cent. of *all* the plaice which were present on the same area must have also been caught. For the marked plaice do not usually suffer from the operation of affixing the label, and they distribute themselves over a wide area. But again the validity of this method has been denied by those whose genius confines itself to the task of criticism. It is contended that the marked plaice segregate themselves and that the percentage caught is more or less accidental and depends on the distribution of the fishing boats. But the advocates of the method claim that the number of the marked fishes recaught is a measure of the intensity of fishing; and the results of such experiments made in England and on the continental side of the North Sea indicate that about 25 per cent., on the average, of all the marked plaice liberated are again caught within the year after the date of liberation. Perhaps it would be straining the results of these experiments to maintain that man annually catches one quarter of all the marketable plaice in the sea, but it is not really improbable that such may be the case.

By far the most satisfactory evidence of the extent of depreciation of a fish population is to be afforded by quantitative plankton investigations such as were carried out by Hensen in the "Nordsee Expedition" of 1895. We cannot, by any means, directly estimate the absolute number of (say) plaice in a sea area such as the North Sea, but we can estimate the number of the eggs spawned by these fishes resident there, and from annual variations in the number of eggs it is possible to estimate the variations in the number of mature spawning plaice. We have seen that this method requires considerable care and must be carried out on a large scale, but it is quite a practicable one, and is, in addition, theoretically quite a sound method. It also affords in an indirect way evidence of the extent of depletion of the stock due to fishing operations. Hensen carried out extensive quantitative plankton observations in the West Baltic and ascertained that in each square metre of sea in this area during the months of January to April, there were produced 370 eggs of plaice and cod.

The fishery statistics of this neighbourhood also shewed that a certain number of mature cod and plaice were annually caught. Then taking the average annual catch of these fishes (deduced from a nine-yearly average) Hensen shewed that about 23,400 million cod eggs, and 73,895 million plaice eggs would have been spawned by these captured fishes if they had been allowed to remain in the sea. When these totals were divided by the total number of square metres in the area fished over, it was found that the fishes caught would have produced 110 eggs per square metre if they had been left in the sea. That is the total productivity of the fish left in the sea and those caught on the area was 480 eggs per square metre.

But the yearly average catch was 110 eggs per square metre, that is about one fourth of the total number theoretically producible. So Hensen concluded (and there does not appear to be any defect in the argument apart from the errors of observation involved in the method) that the fishermen of the West Baltic captured annually about a quarter of the total number of the adult cod and plaice present in the area fished over during the annual spawning season.

Thus we cannot come to any other conclusion than that fishing operations, as at present carried out, do cause a very appreciable diminution of the stock of fish on the sea bottom. Probably some species of fishes, like the herring, mackerel, and haddock, are not appreciably lessened in number by the fishermen, but all that we know shews that this is not the case with other species like the plaice and sole. In the exploitation of a fishing area it is the larger and older fishes of a species which first suffer diminution, and it is the common experience of fishermen that the average size of the species which suffer depletion becomes lowered as the result of intense fishing. We have to determine then what degree of impoverishment such an area as the North Sea will suffer; that is, what is the quantity of fish (like plaice, soles, &c.) that can be annually drawn from it to the greatest advantage. If we take more than this quantity then the stock will gradually become diminished; if we take less then the area is not exploited to the greatest advantage. But this limit of exploitation is probably recognised by the fishermen, for when their operations become less profitable they desert the partially impoverished grounds for others.

The productivity of the sea in different latitudes. In comparing the pictures of life presented by land and sea I supposed that we were exploring a part of the land in a temperate zone. It is well known that the density of life, particularly plant life, on the land is greatest in the tropics and least in the polar regions. The intensity and duration of sunlight is at a maximum in equatorial lands; the temperature, too, is highest. Whether we proceed from the torrid to the frigid zones, or ascend from the sea-level into mountainous regions, we find that both the nature and density of plant and animal life change regularly with the isotherms. Everywhere (except in desert regions) the hotter lands have the most luxuriant vegetation and the colder lands the least. "He who," says Brandt, "has to force a way through the dense plant life of a primeval tropical forest, and then sees the stunted vegetation of Spitzbergen just emerging from the soil, is easily convinced of this contrast."

Naturally one supposes that this is also true of the sea. In tropical regions we find a greater variety of marine animals. The colouring is more vivid and the form is more varied and ornamental. The massive coral and other lime formations due entirely to the activity of coelenterates impress one with a sense of the luxurance of life, and we are apt to forget the factor of time in the building up of these monuments of animal life. Unconsciously one confuses variety of forms with richness of individuals. The usual picture we obtain from records of voyages in tropical seas is that of the wealth of life, and we are naturally surprised to find that this impression is a false one and that the tropical seas are not more abundant in plant and animal life than the temperate and polar waters; and indeed that the reverse is really the case.

There is little doubt that the distribution of life in the sea is exactly opposite to that on the land. The greatest fisheries are those of temperate and arctic seas—these are the White Sea, the Icelandic waters, the Lofoten cod-fisheries, the Newfoundland fisheries, and the rich fishing grounds of the North Sea, and off the north and west coasts of Scotland. The cold seas of arctic and antarctic regions are the chosen haunts of the largest creatures that exist on the earth; and from the earliest recorded times man has sought these inhospitable climes for the whales and seals. Many fishes, such as the sharks and halibuts inhabiting the cold seas, are among the largest of the elasmobranch and teleost orders. Nowhere are sea birds so numerous as in polar waters. The benthic fauna and flora are also most luxuriant. Schimper[1] tells us that the macroscopic vegetation of the tropics is less abundant, and is less rich in species, than that of the temperate seas. The algal flora of the Arctic, though less rich in species, and covering a smaller area than that of the tropics, surpasses in density that of all other seas with the exception of the Antarctic Ocean. Kröyer found an abundance of life such as he had never seen surpassed, and seldom equalled, in Belsund, in Spitzbergen (77° N. Lat.). The sea bottom was, without exaggeration, covered with ascidians and molluscs. The seals and sea birds which he dissected had their stomachs full of crustaceans. Amphipods were so abundant that in one night the carcase of a seal was entirely cleaned and reduced to a skeleton. A basket containing the head of a shark was sunk to 75 fathoms and when hauled in two hours it contained six quarts of amphipods, though as it came up a cloud of these crustacea issued from it like a swarm of bees and were left behind. K. E. von Baer, in 1864, thought it doubtful whether, in the polar seas, life was not more abundant than in the much greater area of the equatorial ocean. One obtains a striking picture of the wealth of marine life from Darwin's account of the biology of Tierra del Fuego[2]. The inhospitable and desolate land contrasted strongly with the sea. "If we turn from the land to the sea," he says, "we shall find the latter as abundantly stocked with living creatures as the former is poorly so....There is one marine production, which from its importance is worthy of a particular history. It is the kelp or

---

[1] *Pflanzengeographie auf physiologischer Grundlage*, Jena, 1898.
[2] *Journal of Researches.*

Macrocystis pyrifera. This plant grows on every rock from low watermark to a great depth, both on the outer coast and within the channels. I believe during the voyages of the *Adventure* and *Beagle* not one rock near the surface was discovered which was not buoyed by this floating weed." It grows up from the depth of 45 fathoms. "The number of living things of all Orders, whose existence intimately depends on the kelp, is wonderful. A great volume might be written describing the inhabitants of one of these beds of sea-weed. Almost all the leaves, excepting those that float on the surface, are so thickly encrusted with corallines as to be of a white colour. We find exquisitely delicate structures, some inhabited by simple hydra-like polypi, others by more organised kinds, and beautiful compound ascidiae. On the leaves also, various patelliform shells, Trochi, uncovered molluscs, and some bivalves, are attached. Innumerable crustacea frequent every part of the plant. On shaking the great entangled roots, a pile of small fish, shells, cuttlefish, crabs of all orders, sea-eggs, starfish, beautiful Holothuria, Planaria, and crawling nereidous animals of a multitude of forms, all fall out together....I can only compare these great aquatic forests of the southern hemisphere with the terrestrial ones in the intertropical regions. Yet if in any country a forest were destroyed, I do not believe nearly so many species of animals would perish as would here from the destruction of the kelp. Amidst the leaves of this plant numerous species of fish live, which nowhere else could find food or shelter; with their destruction the many cormorants and other fishing birds, the otters, seals and porpoises, would soon perish also; and lastly, the Fuegian savage, the miserable lord of this miserable land, would redouble his cannibal feast, decrease in numbers, and perhaps cease to exist."

Just as the algal flora and the benthic and nektic life of the temperate and polar seas are richer than those of the tropical waters, so too the plankton of the colder waters is more abundant, though perhaps less varied, than that of the warmer. The whalebone whales are plankton feeders, and one may well wonder at the huge masses of pteropods that must exist in Arctic seas to afford the necessary amount of food for these enormous creatures. Then the richness of the diatom flora of the polar seas is well known. At times the water is visibly discoloured by these organisms, and we are told that fishing operations in northern seas are sometimes impeded by the rank luxuriance of diatom life. In the Antarctic we have a belt of ooze formed predominantly by diatom shells extending round the southern hemisphere and $10\frac{1}{2}$ million square miles in extent: of course one must not urge this as a necessary proof of the abundance of diatom life in the sea since we do not know the rate of formation. But we know from quantitative plankton researches how abundant these microscopic plants are in Arctic waters. Nowhere did the Kiel planktologists make such rich catches as in the Karajakafjord on the north-west coast of Greenland (70° N. Lat.). In October, 1892, Vanhöffen made a catch of 225 c.c. of plankton with a medium size quantitative net hauled up from 29 metres in depth[1]. Even during the months February—June, plankton was present in the water under the ice. Numerous investigations made by Lohmann, Dahl, Schütt, and Krämer in the tropics shewed that the catches made in these warmer seas were generally poorer than those taken from temperate and polar waters[2]. Just the same result was obtained in the course of the Plankton-Expedition. The reader should consult Schütt's chart of the hauls[3] to see how the quantity of the catch varies inversely as the temperature. Relatively enormous catches were made in the cold water of the West Greenland and Labrador currents, and in the up-welling cool water off the west coast of Africa. But everywhere else (in the warm equatorial streams, and in the Florida current) the catch was much less; and in the Sargasso Sea, where the temperature was uniformly high, over twenty catches were uniformly low and contained the minimum amount of plankton caught during the expedition. Yet who, from purely *à priori* considerations, would not have anticipated that just the opposite results would have been obtained? Again one thinks of the Bay of Naples as the place where good zoologists (like good Americans) go when they die. Here if anywhere one would expect an abundance of life in the sea. But just here we find a richly varied, but (in mass) a scanty fauna and flora. Schütt[4] found that the plankton

[1] "Fauna u. Flora Grönlands," in *Grönlands Exped. des Gesell. d. Erdkunde*, 2. Bd., 1 Theil, Berlin, 1897.
[2] Summarised by Brandt, "Stoffwechsel im Meere," 2 Abhandl. *Wiss. Meeresunt. Kiel. Komm.*, Bd. vi. Abth. Kiel, 1902.
[3] *Analytische Plankton-Studien*, Kiel u. Leipzig, 1892.
[4] *Ibid.*

catches made here were no more abundant than in the Sargasso Sea. Compare this with the Bay of Kiel in the cold Baltic. Though Hensen and Brandt made 70 quantitative hauls during the years 1889—1893, they found on only one occasion (February, 1894) so small a catch as was obtained from the sub-tropical Sargasso Sea[1].

One would naturally expect that the pellucid and warm seas of the tropics would afford the richest fauna and flora; and that the cold and turbid seas of the north would be relatively poor in life. "The yellow Baltic," says Schütt[2], "even in full daylight, only allows us to see the white net at a few metres beneath the surface; the green North Sea transmits light from a much greater depth; and what visitor to the Mediterranean does not know the great transparency of that blue sea? Yet investigations of colour, transparency, and plankton contents give parallel results, and all these shew that the pure blue is the colour of desolation of the high seas."

Thus the colder seas are richer in life than the warmer ones; or at the very least the amount of life in polar seas is not less than in the tropics. We know that intense sunlight and high temperature are favourable to plant life and so these results are at first sight astonishing ones. The density of plant life on the land is apparently a function of these two factors; why then is not this the case also in the sea? One would expect that there would not be such a difference between equatorial and polar seas as between equatorial and polar lands. The range of temperature in the sea is only $33.8°$ C. ($-2.8°$ to $31°$ C.) but on the land the extreme difference is $131.5°$ C. ($65°$ to $-66.5°$ C.). But even with this variation in the temperature of the seas one would expect a corresponding difference in the density of life. Why is the opposite the case? "One stands," says Kjellmann, "as before an insoluble problem when he makes a haul with a tow-net in the Arctic and obtains abundant and strong vegetation, and this at a time when the sea is covered with ice, the temperature is extremely low, and nocturnal gloom predominates even at noon."

[1] "Die Fauna der Östsee," *Verhandl. deutschen zool. Gesell.* Leipzig, 1897.
[2] "Das Pflanzenleben der Hochsee," *Reisebeschreibung Plankt. Exped.* Kiel u. Leipzig, 1892, p. 314.

PART III.

METABOLISM IN THE SEA.

CHAPTER X.

THE CONDITIONS OF LIFE IN THE SEA.

WE see then that it is possible to group together all the organisms of the sea in three great categories—the predatory nektic animals, the sedentary animals and plants living on the sea bottom, and the drifting microscopic life of the plankton. The older methods of marine biological research have shewn how the nature and abundance of the organisms of the fauna and flora of the sea vary with certain physical conditions—with temperature, depth of water, and nature of the sea bottom; and have also shewn that regular seasonal variations occur in the course of the year in any one place. Then oceanographical investigations have made us acquainted with the distribution of oceanic drifts and currents, and with their varying intensity from season to season, and from year to year. The results of the older methods of research have been to shew that there is continual change in the composition of life in the sea—that the total mass of organisms there is in a state of "dynamical equilibrium," for there are numerous factors in operation which set up unceasing change. The temperature changes from season to season with a certain regularity; the intensity of sunlight also varies; the salinity of the sea water undergoes continual small changes; and the distribution and intensity of currents and drifts are not always the same. As these conditions change so do the distribution and abundance of marine life. Probably the total mass of life in the sea remains approximately the same from year to year, but since the total mass of food stuffs in the sea is limited (though very great, of course) it follows that if any change in the life conditions is favourable to the increase in numbers of any one group of organisms it must, of necessity, lead to the decrease in numbers of one or more associated groups.

The quantitative methods of research that we have been discussing afford one of the means of investigating the conditions of this state of equilibrium, for we are led to enquire, in the first place, into the problem of the absolute abundance of life in the sea. Probably we shall never be able accurately to estimate the density of the benthos and nekton, but it is tolerably certain that the improvement of the methods of quantitative plankton investigation will enable us to make very close approximations to the actual density of the organisms comprised in this category; and, by deduction, to that of some groups of the benthos or nekton. That we should become able to do so is essential to success in the attempt to understand the conditions of equilibrium. Then we must be able to investigate the conditions of nutrition of marine organisms; and to trace the effect of changes in physical surroundings on the nature and intensity of these processes. To attain success in this latter object we must know very exactly what are the food substances that are utilised by the marine animals and plants; and how and why the nature and abundance of these change from place to place, and from time to time. Even at the present time a good deal of work has been expended on the attempt to answer these questions and I will try to summarise, in this Chapter, the information at our disposal.

(1) The Nutrition of Marine Organisms.

In the case of the terrestrial organisms we are able to make a broad distinction between the "producers" and the "consumers." The producers are the plants, and the consumers are the animals. We divide the animals into carnivores and herbivores, and into those which make use of both animal and vegetable food. But whether or not an animal feeds upon other animals or upon plants it is the case that its mode of nutrition is such as to compel it to make use of the substances which have been elaborated by other living organisms. The "proximate food-stuffs" upon which an animal lives are the proteids, carbohydrates and fats, which have been built up within the living cells of other animals or plants. Proteids, such as the albumens and allied substances which make up the "flesh" of animals, or the nitrogenous parts of the tissues of plants, are among the most complex of compounds known to modern chemical science. Each molecule is composed of thousands of atoms which are grouped together to form the "building-stones" which by their further combinations form the proteid molecule-complex. These proteid "building-stones" are themselves complex molecules—amido-compounds—which are united together by acidic or basic affinities, are "labile" and are probably in a continual state of chemical dissociation while in the condition of living protoplasm. Carbohydrates, such as starches, sugars or gums are much simpler chemical compounds, and differ further from the proteids in containing no nitrogen. Fats are again relatively simple bodies, and differ from both carbohydrates and proteids in containing much more carbon than either of the above substances—probably the numbers of atoms contained in the molecules of these fats and carbohydrates may be counted by the dozen. Now it is essential for the nutrition of the animals that they should obtain their food from these proximate food-stuffs.

The mode of nutrition of the typical terrestrial plant is fundamentally different from that of the typical animal. While the latter demands organised, and therefore chemically complex, food-stuffs, the former may build up its living substance, and carry on its life processes, by making use of the simplest compounds known to chemistry. Salts of ammonia and nitric acid, carbon dioxide, and water are composed of molecules which contain less than a dozen atoms, nevertheless the plant organism can obtain its nitrogen, carbon, hydrogen and oxygen from such simple compounds; and can build them up into its living tissues. A typical plant obtains its ammonia and nitric acid salts from the ground water by means of its roots, and its carbonic acid and water from the atmosphere by means of its green leaves. From these food-stuffs, and with the aid of the energy intercepted from the sunlight, the plant can build up proteid, carbohydrate and fat. These simple food-stuffs are therefore those of the plant organism[1].

The plants are the producers, since they alone can manufacture organic out of inorganic materials. Borrowing terms from animal physiology we may speak of the characteristic process in the life-activity of the plant as an *anabolic*, or synthetic one. It intercepts the energy of the solar light by means of the chlorophyll contained in its green parts; and by the aid of this energy it can elaborate materials which possess little oxygen; and which, when oxidised or fermented can set free a great amount of energy in the form of heat or otherwise. The characteristic process in the life-activity of an animal is a *katabolic*, or analytic one. It ingests and assimilates the highly complex substances prepared by the plants, and either oxidises or ferments these within its tissues; obtaining, in so doing, the energy provided by the synthetic processes of the plant. Thus (on the land) there is a fundamental distinction between the metabolic processes of animals and plants; and since the latter can alone manufacture living, out of inorganic substance, while the former are able to make use only of the products of the vital activity of some other organism, it follows that the totality of animal life on the land depends on the totality of plant life. If the latter—the producers—were gradually to disappear the former—the consumers—would also disappear.

We will see that this sharp distinction between animals and plants on the land does not hold good in the sea. Nevertheless we may, for the moment, consider that it does, so that we may divide marine organisms into those which make use of organised food-stuffs, and those which can make use of inorganic, or relatively simple, nitrogenous and carbonaceous food substances. This will enable us to distinguish between the organised food-stuffs of the sea, and the ultimate food-stuffs.

---

[1] Plant physiologists make a distinction between the "food" and the "raw materials of the food" of a plant. Ammonia and nitric acid salts, carbon dioxide, and water are the raw materials. The proteids and carbohydrates elaborated in the tissues of the plant itself are the foods.

The plankton as food-stuff. By far the greater number of the nektic animals are carnivores, and we find that there are also a considerable number of benthic animals which feed in a similar manner, while among the plankton there is also a proportion of flesh-eaters. Even when a nektic or benthic animal eats some other animal we find that the *food of the food* is, in the long run, an inhabitant of the plankton; while many fishes, molluscs, crustacea, &c., devour the drifting microscopic life of the sea directly. We then regard the latter as a permanent source of food for other marine animals; and it is important for such an enquiry as this that we should be in possession of reliable estimates of the density of the microscopic life of the sea. But the material for the construction of such estimates hardly as yet exists in the literature. The reader will remember that the catches made with nets of Müllergaze No. 20 do not represent the total contents of the sea in plankton organisms, for a considerable proportion of the latter are so small as to escape through the meshes of nets made of this material. Therefore we cannot use the figures obtained by means of such apparatus for the purpose of making estimates of the amount of organised food-stuffs in the unit volume of sea water, or at the best we can only employ these figures for the construction of minimum values. But, as an illustration of what may be regarded as maximum and ordinary values, we may take the results of two estimations made by Brandt and Lohmann. The former is based on a catch made in Kiel Bay, and the latter on a series of estimations of the total plankton contained in the open Mediterranean, off Syracuse.

The haul described by Brandt[1] is one of the richest ever made in northern seas and we may perhaps regard it as giving us a value for the organised food contents of the sea water, which is very considerably above the average. In it were found:

    3173 millions of diatoms (chiefly *Chaetoceros*);
    500,000 peridinians;
    15,000 copepods.

Its volume was 1385 c.c. and the weight of the contained organisms, dried at 100° C., was 1·06 gramme. Now considering only

[1] " Stoffwechsel im Meeres," *Wiss. Meeresunt. Kiel Komm.* Bd. vi. Abth. Kiel, 1902, p. 71. I refer to this haul, and its " reduction " in Chap. VIII. p. 163.

CH. X] THE CONDITIONS OF LIFE IN THE SEA

the diatoms (which make up the greater part of the catch) we find that the numbers given really underestimate the actual contents of the sea in the region investigated. The area of the mouth of the net was 0·1 sq. metre, and it was lowered down to 20 metres and then hauled up. After correcting for the loss due to the escape of organisms through the pores of the net, and for the filtration loss of the latter, Brandt concludes that the catch must be trebled in order to give the actual contents of the column of water filtered. The amount of dry diatom substance contained in one cubic metre of water was 1·97 grammes, and calculating the amount of proximate food-stuffs contained in this[1], we find that one litre of sea water in the Bay of Kiel, on this occasion, contained:

(1) 0·19 milligramme of proteid,
0·05   ,,   fat,
0·43   ,,   carbohydrate.

Lohmann's figures for the plankton contents of the water of the open sea off Syracuse may be utilised in order to obtain similar values. I have already quoted (in Chapter VIII. p. 166) the description of these catches. The reader will remember that Lohmann made use of various forms of filtering apparatus, and that his figures probably represent the actual contents of the water of the area examined. He found that one cubic metre of the water contained:

2,082,740 protophyta,
325,510 protozoa,
17,415 metazoa,
and   785,000,000 bacteria.

Again using Brandt's tables to reduce these figures, we find that the amount of dry organic matter contained in such a catch was 0·01159 mgr. per litre, and further that the quantities of proximate food-stuffs contained in one litre of the water were:

(2) 0·004 milligramme of proteid,
0·0006   ,,   fat,
0·004   ,,   carbohydrate,

[1] Using Brandt's tables in *Chem. Zusamm. Plankt.*

It will probably be fairly accurate, in the present state of our knowledge, to regard the food contents represented by (1) as a maximum value, and that represented by (2) as a minimum value.

The ultimate food-stuffs in the sea. Bearing in mind the distinction implied in the terms producers and consumers, we see that there is a certain mass of food-stuffs contained in the sea in the form of plankton organisms, and that this represents the nutrition available for the consumers. We have now to consider what are the contents of the sea in the form of inorganic food-salts, that is, the food-stuffs available for the producers. We find that the following substances are the materials that are the essential food-stuffs of plants, or of organisms exhibiting a similar mode of nutrition:

(1) Nitrogen compounds, such as salts of nitrous and nitric acids and ammonia, and possibly amines.

(2) Carbonic acid, which may exist in the air in the form of carbon dioxide, or in simple solution in sea water, or in solution in the form of bicarbonates; and also other carbon compounds.

These two classes of substances supply the plant organisms with the carbon and nitrogen necessary for the building up of their tissues, and for the development of the energy which is manifested in their life-processes.

(3) Phosphoric acid, which is present in the sea in the form of soluble calcium phosphates.

(4) Silica, which also exists in the sea in solution as colloidal silicic acid; and possibly also as particles of clay (aluminium silicate) in suspension, capable of decomposition by the organic matter excreted by plant organisms.

(5) Calcium carbonate, also present in solution.

The three latter classes of substances are utilised by marine organisms. Phosphoric acid is necessary for the elaboration of the nucleo-proteid, which is such a significant constituent of the cell-nucleus. It is also required in the formation of the skeletons of the organisms which possess limy substances. Silica is required for the formation of the hard parts of those protophyta and protozoa which have siliceous skeletons; these are the shells, or frustules, of diatoms; the skeletons of radiolaria; and the spicules of some sponges. Lime, which is usually present in solution as soluble calcium bicarbonate, is required for the formation of the bones of fishes; the shells of molluscs, such as the mussel or oyster; the exo-skeletons of crustacea; the hard parts of corals, echinoderms, polyzoa, some sponges, and some algae; and the skeletons of many protozoa.

(6) Certain other mineral salts, such as the chlorides, sulphates, &c., of sodium, potassium, magnesium, and iron. Only small quantities of these are required.

(7) Oxygen. This is hardly to be regarded as a food-stuff, but it is necessary for the respiration of both animals and plants.

We may now consider the proportions in which these various food-stuffs are present in the waters of the sea.

Nitrogen-compounds. These are present in the sea in excessively small proportion, so small that their determination has always been a matter of great difficulty. Many series of analyses have been made from time to time. The *Challenger* Expedition, which did pioneer work in this, as in so many other directions, collected samples of water from various parts of the sea, and these were analysed by Murray and Irvine some years after the return of the ship. But so rapid has been the progress of oceanographical investigation that we speak of the methods of the last dozen years or so as "modern" in comparison with those of thirty years ago. Our knowledge of the distribution of nitrogen compounds in the sea is due to Natterer[1], Raben[2], and Pütter[3] Natterer investigated the Mediterranean, the Sea of Marmora, and the Red Sea, both at the surface, and at various depths from this to the bottom. He used the best methods of estimation available at the time, and the samples of water collected were examined on board the ship immediately after being taken—an important pre-

---

[1] "Chemische Untersuch. in östl. Mittelmeer.," Ber. Comm. Erforsch. östl. Mittelmeeres. In *Denksch. Akad. Wiss. Wien*, Bde. LIX. LX. LXI. LXII. LXV. 1892–1898.

[2] *Wiss. Meeresunt. Kiel Komm.* Bd. VIII. Abth. Kiel, pp. 83, 277, 1905. See also Brandt, *Rappts. et Proc.-Verb., Cons. Perm. Internat. Explor. Mer.* vol. III. Appdx. D, p. 7, 1905.

[3] "Studien zur vergleichenden Physiologie des Stoffwechsel," *Abhandl. königl. Ges. Wissensch. Göttingen*, Math.-Phys. Klasse, N. F., Bd. VI. No. 1, 1908.

caution, since the presence of marine bacteria soon alters the composition of the nitrogen compounds present. Natterer found that the salts of nitrous and nitric acids were very scarce in the samples of water examined by him—so much so that the quantities just lay on the limits of quantitative expression, and are quoted as "high," "relatively high," "relatively low," "low," and so on. But ammonia salts were more abundant, and were present in measurable quantities. They were most abundant in the water samples taken from the bottom layers, or in the water filtered from samples of mud collected from the bottom deposits. They were least abundant in the upper well-lighted layers of the sea, where plant life was more abundant than at lower depths. If we exclude the results of analyses of bottom samples of water we find that Natterer made 112 estimations of the amount of ammonia salts in the water of the Mediterranean area, with the following results:

87 samples contained 0·0077 to 0·06 milligramme of NH_3 per litre
19 ,, ,, 0·061 ,, 0·1 ,, ,, ,,
5 ,, ,, 0·012 ,, 0·15 ,, ,, ,,
1 sample ,, 0·32 ,, ,, ,,

Natterer's results are apparently unexceptionable and we may take them as representing the nitrogen salts present in the waters of a warm sea area. Raben's results are therefore of particular interest since they refer to the waters of the colder temperate seas. In 1904 water samples were collected, with all possible precautions, by the German International Fishery Investigation steamer, *Poseidon,* in the North Sea and Baltic. The samples were collected in glass tubes previously evacuated of air, and containing mercuric chloride for the sterilisation of the contents. After filtration, so as to remove organisms, Raben estimated the nitrites, nitrates, and ammonia, separately, having previously examined the methods of determining these substances in minute quantity. Nitrous acid was distilled off from the sample, after the addition of acetic acid, and was estimated by means of the metaphenylene-diamine sulphuric acid reaction. Ammonia was distilled off from the sample after addition of magnesium oxide and was estimated, with elaborate precautions, by Nessler's reagent. Nitric acid was finally reduced with sodium amalgam, after precipitation of the mercury salts by magnesium ribbon, and was estimated as ammonia. Raben's results were:

Amount of nitrogen, as ammonia, nitrous and nitric acids, in milligrammes, contained in one litre of sea water.

Date of collection of samples, 1904	No. of samples	N. as NH_3	N. as N_2O_3 and N_2O_5
Open Baltic:			
February	13	0·068	0·199
May	13	0·065	0·170
August	13	0·057	0·095
Open North Sea:			
February	12	0·063	0·216
May	15	0·065	0·217
August	13	0·061	0·079

It will be seen that these results shew a distinct seasonal variation.

Pütter's results are of great interest since they shew that nitrogen compounds, other than those that may be recognised by the methods employed by Natterer and Raben, may exist in solution in sea water. Pütter's analyses relate to samples which were taken from the surface of the Bay of Naples at a distance of three to four kilometres from the shore and they are hardly comparable with those of Natterer, which were collected from the open sea at some considerable distance from the land. Nitrogen compounds could be recovered from the samples by simple distillation in the presence of an alkali, and these were apparently amines and ammonia. In addition to these substances nitrogen was also present as organic matter, which could be estimated by Kjeldahl's method, and it was also present as nitrite and nitrate. Thus a higher nitrogen content was obtained than is indicated by the estimations of Natterer and Raben, but one suspects that some of this excess may be due to the contamination of the sea by discharge from the land, and the insufficient oxidation of the organic nitrogen, through the agency of bacteria. Pütter's results are:

Amount of nitrogen compounds, in milligrammes, contained in one litre of water from the Bay of Naples.

Nitrogen obtained by the Kjeldahl reaction	0·56
Nitrogen as nitrite and nitrate	0·18
Total	0·74

Carbonic acid and other carbon compounds. These are present in the sea in much larger proportion than are the nitrogen compounds. Carbon dioxide is dissolved from the atmosphere; is liberated during the decomposition of organic matter; arises from the respiration of animals; or may be present in combination with bases, as, for instance, in calcium bicarbonate. The amount present per unit volume of water varies greatly geographically, with depth, and with varying salinity of the water; and the whole question of the chemical and physical conditions under which CO_2 may be absorbed by, or liberated from combination in, sea water is much too complex to be discussed here[1]. Pütter, however, made a number of analyses of the amount of carbon dioxide, and other carbon compounds, present in solution in the inshore waters of the Bay of Naples; and his results possess special significance from the point of view of the present enquiry. The methods employed were, apparently, quite sound. The carbon dioxide was liberated from the sample of water by boiling in acid solution in a stream of CO_2-free air, and was absorbed by soda-lime and sodium hydrate and weighed. The other carbon compounds were estimated by Messinger's method—that is, they were oxidised in the water sample itself by potassium dichromate and sulphuric acid, and the CO formed was oxidised to CO_2 by passing over glowing copper oxide in a combustion tube, and was absorbed and weighed. The results thus obtained are tabulated on p. 217. Thus three classes of carbon compounds are present. The nature of the volatile acids and other acids and carbohydrates is uncertain, probably they are substances akin to the humus compounds of the soil, and originate in the mucus, &c., excreted by the algae. Pütter's results are of interest since they indicate that a considerable proportion of carbon compounds may be present in solution

---

[1] See, however, Krümmel, *Handbuch der Oceanographie*, Bd. I. Stuttgart 1907, pp. 303–317.

in the sea, and that these are of such a nature as to serve as the food-stuffs, not only for marine bacteria, but also for some of the lower invertebrates. I believe that these results are quite novel[1].

Amount of carbon compounds, in milligrammes, contained in one litre of sea water from the Bay of Naples.

Carbon compounds	Amount in milligrammes	Carbon in milligrammes
Carbon dioxide	99	27
Volatile acids	36	23
Higher fatty acids and carbohydrates	70	42

Phosphoric acid. This substance is present in the sea in the form of soluble phosphates of calcium. Its exact estimation is attended by particular difficulties, and the older analyses of Schmidt are probably inaccurate. Schmidt[2] found that there were from 2·3 to 5·6 parts per million of calcium phosphate in solution in the warmer seas (Suez Canal, Indian Ocean and South Chinese Seas), and 8·6 to 16·6 parts per million in solution in the colder seas (North Polar, and Norwegian), results not easily to be explained. Raben, however, found that the amount of phosphoric acid in the sea was very much less, and that it underwent a slight, though distinct seasonal variation[3]. His analyses were made on samples of sea water collected by the *Poseidon* during the seasonal cruises, and gave the following results:

Feby. and May, 1904, 0·14 to 0·25 mgrs. per litre of P_2O_5
August   „   1·46   „   „   „

Silica. The amount of dissolved silica in the sea is so small that its exact determination is attended with peculiar difficulties. Here again the older estimations of Murray and Irvine[4] are

---

[1] *Vergl. Phys. Stoffwechsels*, pp. 8–10; and "Der Stoffhaushalt des Meeres," *Zeitschr. Allgem. Phys.* Bd. VII. 2 and 3 Heft, 1907, p. 328–9.
[2] "Hydrologische Untersuchungen," *Bull. Acad. St Petersburg*, Bd. XXIV. 1878. See also Brandt.
[3] See Brandt, *Rappts. Proc.-Verb.* vol. III. Appdx. D, 1905.
[4] *Proc. Roy. Soc. Edinburgh*, vol. XVIII. 1892.

apparently of doubtful value. Silica in solution in water can only be estimated by rendering the substance completely insoluble, and when there is such a large quantity of associated soluble salts as exists in sea water the estimation is very difficult. Raben[1] made a series of analyses of silica in water samples collected in 1902—4 by the *Poseidon* and appears to have obtained results which represent the actual contents of the sea in this substance. In each estimation 3 litres of sea water were evaporated to dryness in a platinum basin. The mass of salts was then treated with hydrochloric acid and again evaporated to dryness, and this process was repeated thrice so as to render the silica completely insoluble. The salts obtained were then drenched with hydrochloric acid, dissolved in water and filtered, and the residue was ignited and weighed. It was regarded as SiO_2, but when it had a trace of colour, indicating the presence of impurity, it was treated with hydrofluoric acid so that the silica could be expelled as silicon tetrafluoride. The residue, if any, was again weighed and the weight deducted from the former value. In this way Raben obtained values which are quoted in the following table:

Amount of SiO_2, in milligrammes, in solution in one litre of sea water.

Date of collection of samples	No. of samples	SiO_2
August, 1902	4	1·037
November „	3	1·26
February, 1903	2	1·45
May „	1	0·65
August „	1	0·93
November „	6	1·084
February, 1904	6	1·015
May „	2	0·655
August „	2	0·926

Thus the proportion of soluble silica in the sea is very minute, at no time exceeding 1½ parts per million of water. We find also that particles of clay are also present in suspension in the shallower

[1] *Loc. cit.* See also Brandt, above.

seas, and Murray and Irvine[1] have made the interesting suggestion that organisms, such as diatoms, are able to decompose these and make use of the colloidal silica set free. Such clay particles may adhere to the film of mucus covering the surface of a diatom, or other organism, and may be decomposed by the sulphur of the organic matter of the excretion of the organism, so affording colloidal silica, which may be used by the latter. But we find that inorganic particles in suspension in river water are rapidly precipitated by the salts of sea water, when they enter the latter, and therefore clay particles are very scarce in the open sea, where nevertheless there may be abundance of diatoms.

Calcium exists in solution in the sea in the form of both sulphates and carbonates; and, relatively to the proportions of silica and phosphates present, may exist in considerable quantities.

Other salts necessary for the nutrition of marine plants are present in the sea in relatively large amount.

Oxygen is present in the sea in proportions that depend on the temperature, salinity and other factors. It is dissolved from the atmosphere, or arises during the metabolism of marine plants. At a given temperature and salinity sea water can only take up a certain quantity of oxygen and this saturation quantity is seldom found in the sea. Indeed in many parts of the latter, where it is imperfectly "ventilated" because of the absence of currents, the oxygen may be greatly deficient, and occasionally its place may be taken by sulphuretted hydrogen. The proportion present therefore varies to a considerable extent, and it would hardly be appropriate here to discuss the results of estimations made in different seas, and at different depths[2]. Pütter found that the water of the Bay of Naples contained, at temperatures of 13° C. to 15° C., an average quantity of about 7·6 milligrammes per litre.

Thus all the substances which are essential for the nutrition of plant life on the land are also present in the sea. We may therefore regard the latter as a dilute food solution which is utilised by the producers, that is, the larger algae, the unicellular algae, the diatoms, and the other protozoa and protophyta which have a similar mode of nutrition. Now it must have occurred to

[1] *Loc. cit.*
[2] See however Krümmel, *loc. cit.* p. 295.

the reader that the proportions in which some of these substances occur, nitrogen compounds, silica, and phosphoric acid, for instance, are very minute, and it may well be asked how such an exceedingly attenuated solution can be utilised as a source of food? It was inconceivable to Murray and Irvine (*loc. cit.*) that diatoms, for instance, could extract the silica necessary for the formation of their shells from solution in sea water, without the expenditure of a considerable amount of energy. We may take it, however, that the difficulty in believing that such a highly dilute solution could profitably be utilised is connected with a natural tendency to regard the metabolic processes in the lower invertebrates, or protozoa (which processes have not been minutely studied) as more or less similar to those of the warm-blooded animals (which have been studied in detail). In order to obtain its nitrogen food from a medium with a similar concentration to that which is utilised by a diatom a mammal would have to pass an enormous quantity of liquid through its alimentary canal, and one can easily see that the labour of such an absorptive process would be very great. But when we consider that such an organism as a diatom, or a flagellate protozoan, probably absorbs its food over its entire surface, and that relatively to its mass this surface is an enormous one, the apparent difficulty disappears. The smaller an organism is, the less food it requires; but then the smaller it is, the greater (relatively) does the surface, by means of which it absorbs its food solution, become. Putting it in mathematical language, we may say that with diminishing size the surface decreases with the square of the radius, while the mass decreases with the cube of the radius. Assuming that the surface of a single bacterium is only $10\mu^2$ we find that one kilogramme of dry organic substance corresponds to a surface of 62,500 square metres! (Pütter[1]). In man, however, one kilo. of dry organic substance corresponds to a surface of only 0·168 sq. metre. One may conclude, then, that food solutions that are totally inadequate for the satisfactory nutrition of a large animal are sufficiently concentrated to be utilised by an organism of the size of a diatom, just because the means of absorption in the latter are so much more powerful than in the case of the former.

[1] "Stoffhaushalt des Meeres," p. 343.

Modes of nutrition among marine organisms. We have distinguished between two great classes of food-stuffs in the sea; (1) those which are contained in the living bodies of animals and plants, and which we have called proteids, carbohydrates and fats; and (2) mineral salts, organic acids, and other relatively simple chemical compounds. Corresponding to this grouping of the food-stuffs we should expect to find a corresponding grouping of organisms into consumers (utilising the organised food-stuffs), and producers (utilising the dissolved inorganic food-stuffs). We do indeed find this division, but the line of demarcation in the sea is not so clear as on the land.

Holozoic organisms. These are they which can utilise as food only the proteids, carbohydrates, and fats, which have already been built up within the bodies of other organisms. The creatures of the sea which exhibit a holozoic mode of nutrition we call animals. All marine mammals, reptiles and fishes feed in this manner, and among the invertebrata the greater number are also predatory animals. As a general rule all animals in the sea exhibit a preference for some particular food-organism, or group of such. Porpoises devour fishes. Most fishes are carnivorous, thus the cod feeds by preference on other fishes and crustacea, though at a pinch they are not fastidious. Whiting, turbot and brill are also fish-eaters, but the haddock seems to prefer invertebrata, such as echinoderms and crustacea. Some flat-fishes, such as the plaice and flounder, feed predominantly upon molluscs, such as the cockle, mussel, and other lamellibranchs (*Tellina, Mactra, Scrobicularia*, &c.), but others, like the dab, are almost omnivorous, and eat anything from a fish to a zoophyte. The sole feeds on worms. Some of the molluscs, like the cuttlefishes and whelks, are carnivorous. So also are most of the crustacea, which are generally garbage eaters, and feed upon the dead bodies of fishes and other animals. This is the case with the crabs, lobsters, and amphipods. Echinoderms, such as the starfish, include many carnivorous species among their classes; thus the starfish may be a most formidable enemy to the mussel and other molluscs. The sea-anemones are also, to some extent, carnivorous animals.

All these are examples of marine animals which are distinctly predatory in their habits, seeking and devouring other fairly large animals. A further category of carnivorous animals in the sea includes those which feed upon the plankton, and which possess special organs of alimentation for the capture of these microscopic organisms. The whalebone whales feed upon the pelagic pteropods, and the baleen strainers in the mouth form a filtering arrangement wherewith the animal separates the food organisms from the water taken into the mouth. Fishes, such as the herring, sprat, mackerel, shads, and many others, feed upon the pelagic copepoda and schizopods; and these animals possess gill-rakers, which are fine comb-like structures placed on the internal margins of the gill-bars, and which form a straining apparatus which separates the micro-crustacea from the water taken into the mouth. Many molluscs eat the micro-plankton, which they obtain from the sea water by causing the latter to pass through the fine interstices of the gills, thereby filtering out the particles of food which are then taken into the mouth by means of ciliary action. It is probable that very many other animals are plankton eaters, thus the micro-crustacea doubtless feed in this manner. The molluscs, like the mussels and cockles, which feed on the plankton, are herbivores, since they appear to subsist chiefly on the diatoms. Some fishes, of which the grey mullet is the best known example, eat the finer algae, and the diatoms with which the latter are usually associated. These fishes "browse" upon the sea-weeds growing on stones, &c., in much the same way as a cow eats herbage. Other fishes, like the sardine, also feed upon diatoms, but they obtain these in the same way as the herring obtains the copepods and schizopods on which it subsists.

Thus there are marine animals which eat the organisms of the plankton. But if we push our enquiry into the food organisms of sea animals far enough we find that the plankton creatures are the last links of chains of food organisms. Thus the cod feeds on plaice, which feed on molluscs, which feed on diatoms, and other protophyta. The plaice is a food organism of the first degree, the mollusc one of the second degree, and the diatom one of the third degree. Generally speaking we find that the food organisms of the nth degree of a carnivorous animal are such as belong to the protophyta of the plankton. And this is why we speak of the plankton as constituting the ultimate organised food-stuff of the sea.

In all these cases the metabolic processes concerned in nutri-

CH. X]   THE CONDITIONS OF LIFE IN THE SEA   223

tion are essentially similar to those which are carried on in the warm-blooded animal. The "raw materials of the food-stuffs" are (1) the proteids of the food organism. These are digested in the alimentary canal, and the soluble peptones so formed are absorbed by the intestinal walls, and, in this process, are converted into the proteid which is characteristic of the animal concerned. (2) The carbohydrates, which are hydrolysed during digestion so as to form diffusible sugars which are then absorbed by the intestinal walls. (3) The fats, which are dissociated in digestion and are again synthesised in the process of absorption. The undigested, or undigestible residue of the food is expelled from the alimentary canal as the faeces. The products of digestion, that is, the reconverted specific proteids, the sugars, and the reconverted fats are carried in the blood stream to the tissues, there to be built up into living substance, or to form stores of reserve materials; and ultimately to undergo combustion by the inspired oxygen, yielding in this process the energy of the animal. Some of this expended tissue substance is excreted in a completely oxidised form as carbon dioxide and water, and the remainder as the incompletely oxidised products, urea, uric acid, guanin or hippuric acid. So far we are on familiar ground, and processes, characteristic of the metabolism of the warm-blooded animal, are carried on also in the life economy of the higher marine animals.

Holophytic organisms. In complete contrast with these processes stand those which are generally characteristic of the higher plants. Here the "raw materials of the food-stuffs" are the mineral salts, ammonium sulphate and carbonate, alkaline nitrates and the carbonic acid of the sea water. The typical plant builds up starch from the simple compounds carbonic dioxide and water, and amido-substances from the mineral salts taken up by its roots. From these substances it prepares the proteid of its tissues. This extraordinary power of forming starch is called photo-synthesis, and it depends on the existence, in the green parts of the plant, of the pigment, chlorophyll. This substance can intercept the energy of sunlight and utilise it in the elaboration of starch.

Carbon dioxide and water, the two compounds from which the starch is built up, are completely oxidised substances. But starch is a compound which contains carbon capable of further oxidation, and therefore in the synthesis of the latter compound from water and carbon dioxide oxygen must be set free. So we find that a green plant when living in the light, and assimilating the carbon of its medium, gives off elementary oxygen. But it also respires in the same manner as an animal, that is, it takes in oxygen, which burns up some of its tissue substance, giving off carbonic acid. In sunlight the amount of oxygen which is given off in the process of carbon assimilation is greater than the amount of the same element which is used up in the process of respiration. Therefore we say that a green plant gives off a surplus of oxygen to the atmosphere, or to the sea water if it is a marine species. But in the dark, when the stimulus of the radiant energy of sunlight is withdrawn, the process of photo-synthesis ceases and the plant then gives off only carbonic acid, but continues to take up oxygen just as an animal always does.

The green, red, and brown sea-weeds of the shallow water; the pelagic algae the diatoms; and the other protophyta of the plankton are the holophytic plants of the sea. But the larger algae are restricted to a comparatively narrow well-lighted zone of sea-bottom near the shore, while the protophyta are present everywhere in the sea where sufficient light penetrates. It is the latter microscopic plants therefore which are responsible for the greater part of the production in the sea.

Saprophytic organisms. Many of the lower plants grow luxuriantly on soils, or media, containing organic matter, such as sugars or organic acids. These saprophytic plants are represented chiefly by the moulds, yeasts, and fungi. Their food-stuffs are therefore more complex in chemical structure than those of the holophytic plants, and their energy is obtained not so much by the oxidation of the tissues of their bodies as by the fermentation of the materials on which they live. These food-stuffs contain much energy and in the process of fermentation they are decomposed, and a small part only of the energy thus liberated is used up by the saprophyte. A typical case is that of the fermentation of a sugar in the process of brewing. Here the latter substance is decomposed with the formation of alcohol and carbon dioxide, and a part of the energy thus set free is used up by the yeast and reappears as the vital activity of that organism. The yeast multiplies and in its metabolism the enzymes, or ferments, which are instrumental in the splitting up of the sugar to form alcohol and carbon dioxide, are excreted. The saprophytic plants in the sea are moulds, yeasts and fungi, but they are not abundant. The marine bacteria also exhibit a mode of nutrition which is related to that just discussed, but it will be better that we should consider it separately.

Myxotrophic animals. We can only draw a rather indefinite line of demarcation between plants and animals. In the case of the higher members of each kingdom there is, of course, a very sharp distinction, both in structure and life-history, and in the general character of the metabolism of the organism. But among the lower animals and plants this distinction disappears and several groups of organisms have at one time been included among the plants, and at another among the animals. Certain well-known characters are distinctive of the plant—the relatively thick cell-wall of cellulose; the presence of chlorophyll bodies in the green parts; and the general absence of motility. On the other hand the much less strongly developed cell-wall; the general presence of chitin instead of cellulose; the absence of chlorophyll; and the general presence of organs of locomotion, are characteristic of the animal. But we find that investments composed of a substance resembling cellulose may occur among animals, while the cell-walls of bacteria appear to be composed of chitin[1]. Chlorophyll bodies are absent in many plants, and are, apparently, present in many animals. Diatoms, algal spores, and bacteria are actively motile, while some animals, such as the trophic forms of myxosporidia, are immotile. There are many species among the flagellate protozoa which, if we were to consider the morphology alone, we should call animals; while their mode of nutrition indicates that they belong to the vegetable kingdom. All this confusion need not, however, trouble us. In the sea there are only individuals and whether we call them animals or plants is not of importance in our enquiry. Morphological distinctions, it has been said, are ultimately physiological ones; and, as we are beginning to learn, depend on the almost infinite diversity of conformation of the proteid molecule-complexes which make up their protoplasm.

[1] Mary Leach, *Journ. Biol. Chem.*, 1906, (1), p. 463.

A still further complication is encountered in the cases of those animals which contain green or yellow cells in their tissues. Here we have to deal with associations of two organisms—the animal and the contained green cells, which belong to one of two groups of algae—the Xanthellae and Zoochlorellae. These partnerships arise from the "infection" of the animals by the cells of the plant, and by the further multiplication of the latter within the tissues of the animal; and we speak of such associations as "symbioses," or sometimes as cases of parasitism. The most familiar example of these associations is the well-known green Hydra, the tissues of which contain green corpuscles, which have long been regarded as "commensal algae." Many other cases are to be found among the Protozoa, Sponges, Corals, Mollusca, Alcyonaria, Hydrozoa, Medusae and Turbellaria; and even among the much higher groups of the Mollusca, Polyzoa and Echinoderms. Sometimes the association is an obligatory one, that is, neither of the partners can exist without the other, or at least, one of them must be associated with the other in order to live. Sometimes, more often indeed, the association is a facultative one, that is, the animal can exist apart from the alga, and *vice versa*.

Now the point of interest for us in connection with these associations is that two modes of nutrition may proceed simultaneously in the same (compound) organism. Indeed the question of how the thing feeds itself is often an extremely complicated one. The most exhaustive investigation of the life-history and metabolism of such a symbiotic organism is that made by Keeble and Gamble[1] in the case of the Turbellarian worm *Convoluta roscoffensis*. This animal begins life as a colourless creature, and within three days after birth becomes infected with the cells of a species of alga which lives saprophytically in the water in which the worm also lives. This alga belongs to the group Chlamydomonadeae. It multiplies in the tissues of the worm and soon the latter becomes green, the colour being due of course, to the contained algal cells. The plant may exist outside the tissues of the animal, when it possesses the form typical of its group, but when the infection has taken place the cells undergo

[1] "The Green cells of Convoluta roscoffensis," *Quart. Journ. Micro. Science*, vol. LI. p. 167, 1907.

a certain degree of degeneration, except that the photo-synthetic machinery and functions remain intact and unimpaired.

What this means is that the green algal cells are able to intercept the energy of sunlight, by virtue of the chlorophyll that they contain; so that they can build up starch from the carbon dioxide and water of the medium in which they live. They obtain their nitrogen from the waste nitrogenous products of the metabolism of the body of the worm. Thus the mode of nutrition of the green algal cells is holophytic, but they also betray a tendency towards a saprophytic habit in that they can utilise the nitrogen of such complex substances as urea, or uric acid.

On the other hand the worm begins life as a typical animal; it captures and ingests organisms like diatoms, and this mode of nutrition—a holozoic one—continues for about a week after birth. If it does not become infected then with the alga it dies, probably because it is unable to get rid of the nitrogenous waste products of its own metabolism. The infection however always becomes established in the state of nature, and then the excretory products of the *Convoluta* are made use of as food by the algal cells, while the starch elaborated by the latter is converted into sugar and is used as food by the worm. Thus the association is a mutually beneficial one for both organisms.

We encounter similar cases of symbiosis among the corals and coelenterates[1]. It is well known that much difficulty is experienced in finding food materials in the digestive cavities of tropical corals in sufficient quantity to serve for their proper nutrition. Some time ago both Brandt and Hickson suggested that the food-stuffs of these animals were supplied to them by the Zoochlorellae which are often contained in their tissues. In the paper by Miss Pratt, to which I refer, satisfactory evidence is furnished that this is really the fact. The Alcyonaria of British seas do indeed capture and ingest living prey, such as copepods; paralysing the latter by means of the poison threads of their nematocysts, but these animals do not contain green algal cells. In tropical Alcyonaria we find, however, that infection by Zoochlorellae takes place, and that the increase of the infection is

[1] Miss Edith Pratt, "Digestive organs of the Alcyonaria," *Quart. Journ. Micro. Science*, vol. XLIX. p. 327, 1906.

accompanied by the progressive decrease of the digestive surface of the autozooids. With this reduction we find that the animals have lost the power of capturing and digesting living food. Among the true corals (Madreporaria) the same association may be seen; that is, green cells, with contained chlorophyll and starch granules, are contained in the tissues of the soft parts of the corals, and these green cells are most abundant in those parts of the coral which are exposed to the light, and are in continual contact with the current of water which passes through the cavities of the animal. The green cells are included Zoochlorellae, and they take up carbonic acid from the sea water in which they live, and by virtue of the power of photo-synthesis which they possess, elaborate starch, which is used by the coral in the same way as in the case of the green *Convoluta*.

Saprozoic animals. Such associations between animals and plants—symbiosis, or parasitism, whatever name we give them— do not really invalidate the distinction between the characteristic holophytic mode of nutrition of the plants, and the characteristic holozoic mode of nutrition of animals. We can still clearly distinguish between the two members of the association, and their distinct modes of nutrition. But the existence of a saprozoic mode of nutrition among animals deprives the method of feeding, considered as a means of distinguishing between the two kingdoms of life, of its value. A saprozoic animal resembles precisely a saprophytic plant, for both find their nutritive material in complex chemical substances, such as peptones or extractives, or carbohydrates and organic acids. Still further we find that there are apparent instances of photo-chemical reactions in the metabolic processes of some animals.

Many of the protozoa are saprozoic in their mode of feeding. If we add some sugar to a little water containing bacteria we will find that very soon there will arise an abundant infusorial growth. Very probably the infusoria feed upon the added carbohydrate. All internal parasites are also saprozoic. A tape-worm possesses neither mouth nor alimentary canal, and it lives throughout its life (except perhaps for a very short period, when it may inhabit water or other media) attached by means of suckers or hooks to the wall of some internal cavity of an animal body.

The whole surface of the skin of an internal parasite is bathed in the juices of the host—the blood, if it lives in the lumen of the heart, or a blood-vessel; the lymph if it inhabits the peritoneal cavity; or the digested food-substance if it lives in the intestine. All these liquids are rich food media, and the parasite simply absorbs the nutriment which they contain over the entire surface of its body. We usually make, or imply, some distinction between the method of feeding of a parasite and that of a closely allied species; most biologists would assume, for instance, that the mode of nutrition of an *Ascaris* or *Bilharzia* was different from that of the free-swimming stage of a *Gordius*, and that a parasitic Turbellarian like *Graffilla* differed in a corresponding manner from those species which live in the open.

Comparative physiology has not received anything like the same amount of study as comparative anatomy; and so we find that our notions of the metabolism of the lower invertebrata has been profoundly influenced by our extensive knowledge of the physiology of the warm-blooded animals. Morphological studies have enabled us to trace homologous structures and organs through long series of animal forms; thus the ear-ossicles, and parts of the jawbones of the mammal, are homologous with parts of the visceral skeleton of the fishes; and the lungs of the warm-blooded animal are morphologically the same thing as the swim bladder of the fish. But whereas in these specific cases we recognise quite clearly that the functions of the homologous structures are different, we do not always bear in mind that the functions of morphologically similar structures, in the invertebrata, may be entirely different. So great is the force of well-known words that we naturally assume that the gills of a fish, the gills (ctenidia) of a mollusc, and the gills (respiratory plumes) of a tube-living worm have the same functions, those of taking up oxygen from, and giving off carbonic acid to the sea water. Just in the same manner we assume that the digestive cavity (stomach and intestine) of a fish, the digestive cavity (coelenteron) of a coral polyp, and the digestive cavity (gastric pouches and canals) of a medusa, have similar functions; those of digesting and absorbing the materials of the food-animals captured.

So also with regard to our notions of the nature of the

metabolism of the lower marine animals. Without sufficient reflection we assume that the oxidation process characteristic of the mammal also obtains in all animals living in the sea. We naturally assume that the proteid, carbohydrate and fats of the bodies of animals are ingested, converted into peptones, sugars and dissociated fats, are absorbed and built up into tissue; and that the constituents of the latter are then oxidised by the oxygen taken up by the gills and excreted as carbon dioxide, water, and nitrogenous waste products like urea or uric acid.

Do marine animals always obtain their food supply by ingesting the bodies of other animals and plants? One naturally assumes that this is universally the case and so encounters difficulties, such as I have often experienced in attempting to demonstrate the food of a cockle to a class of sceptical fishermen. Repeatedly one searches among the mud and sand that fill the alimentary canal of this mollusc for traces of planktonic organisms, and it is only the magnifying power of a $\frac{1}{6}$th inch objective, when applied to one or two *Naviculae*, that makes the demonstration a partial success. It is often with great difficulty that one succeeds in finding food in the alimentary canal of some invertebrates. Dohrn failed to demonstrate the food of Pycnogonids, and Rauschenplatz, in an extensive study of the foods of marine invertebrates of the Bay of Kiel failed, in some cases, to find satisfactory evidence of the food organisms[1]. We can hardly ever find visible food within the alimentary canal of a mature plaice during the winter months, and yet it is just during these months that the reproductive organs of the fish are maturing for the spawning act in the following spring. Often one may observe that fishes and other animals may live for a long time in aquaria without obvious food. Do they really cease to feed during this time; and if so why is the plentiful supply of oxygen in the water conveyed to them quite essential?

Pütter seeks to answer these questions by suggesting that the sea is an immense reservoir of food-stuffs in the form of dissolved organic carbon and nitrogen compounds. We have already seen that there is evidence that these substances exist in the sea— at least in the water of the Bay of Naples. He suggests that

[1] See also Shipley, "Gephyrea," *Cambridge Nat. Hist.*, vol. 2.

many marine animals habitually feed on this dissolved food-stuff, behaving, to that extent, as saprozoic creatures. Pütter investigated the metabolism of several marine organisms by estimating the amounts of oxygen, carbon and nitrogen exchanged between the organism and its medium in the unit of time, and per unit of weight. In the case of the sponge, *Suberites domuncula* he found that an animal of about 60 grammes weight required per hour 0·92 milligramme of carbon[1]. If we assume that the sponge obtains its food by eating plankton organisms then we can calculate the volume of water necessary for the provision of this weight of carbon. Assuming that Lohmann's estimates of the density of plankton in the open Mediterranean are true also of the Bay of Naples, Pütter found that it would be necessary for the sponge to capture, *per hour*, all the plankton contained in 242 litres of sea water, that is about 4,000 times its own volume!

This was obviously impossible, for a rough estimation of the volume of water passing through the osculum of the creature shewed that this could not be more than about 300 cubic centimetres; and this volume of water contains only $\frac{1}{810}$th part of the carbon required by the sponge. But if the reader will refer to Pütter's estimate of the amount of carbon compounds contained in the sea he will find that only 14·2 c.c. contained as much of this element as the creature demanded. So even if we assume that all the carbon compounds contained in the water are not capable of utilisation as food, and that the absorption coefficient is not a very high one, it is still the case that the water circulated through the cavities of *Suberites* may have contained enough carbon to supply the hourly requirements of the animal.

Pütter also found that the Holothurian, *Cucumaria grubei*, when kept in an aquarium in filtered sea water, lost weight. Now calculating the loss of carbon due to the loss of body weight he found that the average deficit was 0·015 mgr. per hour for a confinement of six months. But by estimating the actual amount of carbon given off in the metabolism of the animal he found that this was 0·2 mgr. per hour. Thus more than $\frac{9}{10}$ths of the carbon exchanged between the animal and its medium could only have

[1] "Die Ernährung des Wassertiere," *Zeitschr. Allgem. Phys.* Bd. VII. 1907, p. 292.

come from carbon compounds present in solution in the water circulating through the aquarium[1].

Is a large part of the carbonaceous food of these animals (and by implication, a part also of the nitrogenous food) obtained from compounds in solution in the sea? I think that the evidence outlined above, though indirect, is fairly conclusive in favour of such a statement. If it is, what is the nature of the metabolic process to which this dissolved food-stuff is subjected in its passage through the tissues of the animal body? This cannot be similar to the oxidation process which is characteristic of the warm-blooded animal. In the case of the latter the absorption of oxygen from the atmosphere is carried out in the lungs, and probably more of the gas could be taken into the blood stream than would be sufficient for the oxidation of the food products in the tissues. But this is probably not the case with the two animals studied by Pütter. The following figures show the results of several estimations of the CO_2 and O exchanged[2]:

Respiratory exchange in Suberites *and* Cucumaria.

Temperature	CO_2 given off	O taken in	Respiratory quotient
Suberites			
13·5° C.	282 mgrs.	30 mgrs.	6·5
Cucumaria			
11·7° C.	140 ,,	27 ,,	3·8
13·7° C.	166 ,,	33 ,,	3·68
17·0° C.	211 ,,	60 ,,	2·55

These results show that the amount of oxygen contained in the carbon dioxide given off is far more than can be accounted for by the oxygen taken in; that is to say the respiratory quotients are astonishingly high. All the carbon dioxide which is given off cannot proceed from the oxidation of carbon, since not enough oxygen is taken in to balance the gas contained in the CO_2 which is excreted. In the case of the fermentation of sugar by a yeast, both carbon dioxide and alcohol are formed but the former does

[1] *Vergleich. Phys. Stoffw.* p. 61.
[2] *Ibid.* pp. 33 and 52.

not originate by a process of combustion but by the splitting up of the carbohydrate. Doubtless this is, to some extent at least, the case also in the metabolic processes which we have just been discussing, that is the sponge and cucumarian obtain a part of their energy by making use of a process of splitting of the carbon compounds absorbed. Of course oxidation does also take place, but it is subordinated to the fermentation reactions.

In the case of the sponge we can easily see that there may be a considerable amount of absorption of dissolved food-stuffs in the process of alimentation. In these organisms there is hardly any differentiation of the tissues into organs of digestion, respiration, and so on; and the whole internal surface of the body of the animal must function as an organ of food absorption and respiration. But in the case of a holothurian there is a well-marked alimentary canal and respiratory organs. So also in the case of such organisms as molluscs and ascidians, and one may ask, what are the uses of these organs if the animals feed by absorbing dissolved nutriment, and if the respiratory function is much less important than it is in the case of a warm-blooded animal? The ctenidia of a mollusc, and the gills of an ascidian, are usually assumed to be respiratory organs, but it is quite certain that a very important function of these structures is that of setting up the current of water which enters into, and passes through the cavities of the animal's body. Now if it is the case that the oxidation process is less important in the metabolism of these animals than is the fermentation reaction, we encounter a difficulty, for why should such an abundant supply of oxygen (in the water current) be necessary to an animal which apparently requires less of this element than those higher in the scale, and which is almost entirely sedentary in its habits. Since the so-called respiratory surface of the mollusc or ascidian is a very large one we may conjecture that it has some other purpose; and Pütter's suggestion that it is a surface for the absorption of dissolved food-stuffs appears to be a probable one. At any rate the actual evidence that the gills of these creatures are organs which function in the absorption of dissolved food-stuff is just as strong (or as weak) as the actual evidence that they take up dissolved oxygen from solution in the sea water.

What then is the function of the alimentary canal in such animals as these? Pütter suggests that the surface of the latter is one which is instrumental in the absorption of dissolved food matter from the sea. That it should also take in and digest solid food-particles, such as the organisms of the plankton, may be a secondary function acquired after the alimentary canal has been evolved for the absorption of liquid food-stuff. Thus the capture and assimilation of diatoms by a mussel or a holothurian may be compared with the capture and assimilation of insects by an insectivorous plant; which latter process we may regard as being strictly secondary in importance to the process of feeding by means of photo-synthesis of starch by the green parts of the plant. In many micro-crustacea we find that pumping movements of the intestine are regularly carried on, water being taken in, and expelled *per anum*. This is usually called "anal respiration," it is just as likely (and it is far more probable *à priori*) that the process represents the circulation of water through the intestine in order that the surface of the latter may absorb the dissolved food-stuffs which are contained in the water.

(2) The Law of the Minimum.

We are indebted to Justus von Liebig for the statement of the "Law of the Minimum." A plant requires a certain number of food-stuffs if it is to continue to live and grow, and each of these food-substances must be present in a certain proportion. If one of them is absent the plant will die; and if it is present in minimal proportion the growth will also be minimal. This will be the case no matter how abundant the other food-stuffs may be. Thus the growth of a plant is dependent on the amount of the food-stuff which is presented to it in minimum quantity. We have seen that marine plants require certain things—carbonic acid, nitrogen compounds, silica, phosphoric acid, and certain mineral salts. The carbonic acid and the mineral salts are present in relatively large amount, but the proportions of nitrogen compounds, silica, and phosphoric acid in the water of the sea are very small. The density of the marine plants will therefore fluctuate according to the proportions of these indispensable food-stuffs.

But we have also seen that the marine animals generally feed on other animals which are smaller than themselves, and these latter upon others which are smaller still, so that the ultimate organised food in the sea consists of the organisms of the plankton. Further it is only the protophyta of the latter that are able to utilise the inorganic salts of the sea water as food; converting these into organic material with the assistance of the solar energy. The abundance of life in the sea then depends on the abundance of the plankton; and the abundance of the latter is dependent on the amount of the indispensable food-salts which are at the disposal of the protophyta. One or more of these substances must be present in minimal proportion, and so must rule the production of organised material in the sea. What is this substance?

It is probable that the abundance of nitrogen compounds in the sea determines the production. Most investigations that have been made shew that there is a relation between the density of the plankton and the proportion of nitrogen compounds. Natterer's estimations shew that the water of the warmer seas is relatively poor in ammonia, and the quantitative plankton investigations of Lohmann, Schütt, and others indicate that these seas are also poor in plankton. Raben's analyses shew that the water of the North Sea and Baltic contains rather more nitrogen than the water of the Mediterranean area, and we also find that the water of these seas is relatively rich in plankton. Apstein[1] made numerous estimations of density of microscopic life in the fresh water lakes of Holstein, and Brandt[2] shewed that those lakes which possessed the richest plankton also had the greatest amount of inorganic nitrogen. Littoral sea water into which flow rivers rich in the nitrogenous drainage of the land, has, as a rule, a richer benthic fauna than the sea further from the shore. Nowhere do mussels grow so luxuriantly as in estuarine waters rich in sewage matters. Fresh water ponds in which carp are cultivated produce varying "crops," but whether the conditions are good or bad the deliberate addition of nitrogenous drainage makes them better.

It is rather difficult to determine which of the three food-

[1] *Süsswasser Plankton*, Kiel and Leipzig, 1896.
[2] "Stoffwechsel im Meeres."

stuffs, nitrogen compounds, phosphoric acid, or silica, is present in the sea in minimum proportion, and so rules the production. It is possible to ascertain the ratios in which the elements, nitrogen, phosphorus, and silicon, exist in the tissues of the more abundant plankton organisms, and Brandt[1] did so by making chemical estimations of the composition of catches which consisted predominantly of certain organisms, such as copepods, peridinians and diatoms. It should be possible, from a consideration of the results of such analyses, to ascertain in what proportions these different organisms require the three elements, for the demand for each will be proportional to the amount of each element present in the tissues of the organism (when due account is taken of the different absorption coefficients of the salts containing the elements). Now from such analyses Brandt[2] came to the conclusion that it was the nitrogen which was present in minimum proportion, but since it was the older estimations of the proportions of silica and phosphoric acid in sea water that he considered, it is probable that his conclusion must be revised. Thus the dry substance of diatoms contains about $1.8\,^0/_0$ of nitrogen, and about $54.5\,^0/_0$ of SiO_2 shewing that about thirty times as much silica as nitrogen is required by these organisms. Yet the greatest amount of the latter substance found in the sea by Raben was only 1·4 mgrms. per litre, while nitrogen was present to the extent of about 0·2 mgrm. per litre. Thus silica was only about seven times as abundant in the sea as nitrogen, though thirty times as much was required. Therefore the silica was present in minimum proportion and not the nitrogen.

So far as the diatoms are concerned it would appear that it is the proportion of silica in the sea water that determines the production. Raben's results are so instructive in this connection that I have represented them in the diagram on p. 237, along with the curve of seasonal variation of diatoms in Kiel Bay. It will be noticed that the plankton curve relates to the year 1892, while the silica curve is for the period 1902—4. But a consideration of the Kiel plankton investigations shews that the form of the

[1] *Chemische Zusamm. Planktons.*
[2] In "Stoffwechsel im Meeres," 2nd Abhandlung, *Wiss. Meeresunt. Kiel Komm.* Bd. VI. Abth. Kiel, 1902.

curve is much the same for the whole series of years covered, and we may take it that it is a general one. All organisms taken in the quantitative nets are included in the results of these catches,

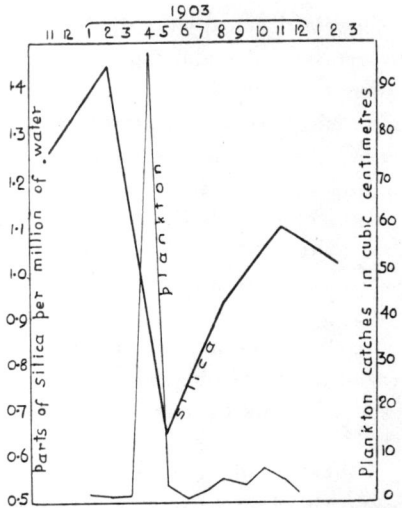

Fig. 30. Variation of plankton and silicic acid in Kiel Bay 1902—4.

but the diatoms form so overwhelming a proportion of the total volume of plankton during the maximal months that we may regard the seasonal variations as due almost entirely to the variations in abundance of these organisms. The thin line then represents the change in the abundance of the plants from month to month, and the thick line represents the variation in the amount of silica in solution in the sea from quarter to quarter.

Note that there are two maxima in the abundance of silica as well as in the abundance of diatoms. At the beginning of the year the proportion of silica contained in the sea is small, and so also are the numbers of diatoms. But the silica gradually increases in amount until the beginning of February, and then, almost immediately afterwards, follows the spring maximum in the abundance of diatoms. Thereupon the proportion of silica rapidly falls, indicating that it has been used up by the plants; and with the fall in abundance of this food-stuff, there occurs a rapid drop in the number of diatoms present per unit volume of water. After the spring months the amount of the silica again begins to increase, and attains a secondary maximum in the autumn, and immediately after this there is a secondary maximum in the amount of the diatoms.

So also with the peridinians, though the relation between the abundance of these organisms and that of the various food-salts, has not been studied in a very satisfactory manner. Still Raben's estimations seem to shew that there may be a relation between the peridinians and the phosphoric acid in the sea, and that the maximum period of reproduction of these protozoa in the autumn may be connected with the maximum period of abundance of phosphoric acid which also occurs in the autumn. But much work remains to be done before this can satisfactorily be established.

Thus there is a fair amount of evidence to shew that the variations in the abundance of the organisms of the plankton are to be attributed to variations in the abundance of the food-stuffs which stand at their disposal; and this is indeed what we should expect. Even in a slowly multiplying human society we can trace a connection between the fluctuations in the labour market (which are indicative of changes in the general prosperity of the artisan class), and minor fluctuations in the marriage and birth rates. An increase in the demand for labour is followed by an increase in the number of marriages[1]. But in the sea the lower organisms multiply very rapidly, and so an increase in the amount of food is soon followed by the multiplication of those creatures which utilise this food. But the greater amount of mouths to be filled soon reduces the available amount of food, and a drop in the rate of multiplication very soon follows.

It does not appear to me that the existence of saprozoic and saprophytic modes of nutrition among marine organisms invalidates the conclusion that the abundance of life in the sea depends on the abundance of those food-stuffs which are present in minimal quantity. If there is an immense store of food in the sea, in the form of dissolved carbon compounds other than carbonic acid, it is still the case that these substances (which may be compared with the humus of the soils) are produced by the metabolism

[1] See *Elements of Statistics*, A. L. Bowley, London, 1902.

of some marine organisms. Probably most of the plants and animals in the sea excrete this dissolved food-stuff (the mucus of algae and of some fishes, like the hag, or skate, for instance); and if we admit that the intensity of metabolism is proportional to the surface of an organism, we may determine roughly what part is played by the various groups of marine organisms. Pütter gives some figures for the construction of such estimates[1]. Assuming a composition of the plankton, such as is represented by Lohmann's tables, and regarding the total surface of all planktonic organisms as equal to 1000, he finds that this surface is shared in the following manner:

>   Protozoa,   29,
>   Metazoa,   148,
>   Bacteria,   400,
>   Protophyta, 423,

and it would appear from these figures that the protophyta (the diatoms and lower algae) play a more important part in the production of dissolved carbon food-stuffs than any other group of organisms in the sea. It is true that the total effect of the bacteria would appear to be almost as great as that of the protophyta, but we shall see that the general effect of the metabolism of the bacteria is to convert all carbon compounds into CO_2, and all nitrogen compounds into nitric acid, or even free nitrogen. It is only the protophyta among the plankton which can utilise the CO_2 and nitric acid compounds, and so we see that upon these rests the greater part of the task of elaborating the organised food-stuff, as well as that of elaborating the dissolved food-stuff of the sea.

(3) Physical Conditions and the Metabolism of Marine Organisms.

Thus we are driven to consider that the variations in the amount of food-stuffs—organic or inorganic—in the sea are factors of the first importance in determining the variations in the abundance of animal and vegetable life, from place to place, and

[1] *Stoffhaushalt des Meeres*, p. 366.

from time to time. But further investigation will shew that certain physical factors—the temperature, degree of concentration of the salts of sea water, and the intensity of sunlight—are also important factors. Hydrography—the study of the physics of the waters of the sea—is now an important department of general marine biology; and the results of this branch of marine research must be considered in all discussions of the question of the variations in the nature and abundance of life in the sea. It is astonishing that it should have been very difficult to induce marine zoologists to take the view that the distribution of species of animals and plants, is in the long run the effect of physical factors; and that hydrographical research is necessary if we are to learn what are the causes of the variations in abundance, and the migrations of marine organisms. One is so accustomed to think of corresponding biological phenomena on the land as dependent on climatic conditions that it is extraordinary that it should be difficult to think similarly with regard to the sea. Mammals, reptiles, some insects, worms and so on live on the surface of the land, at the bottom of the atmosphere, just as benthic animals live at the bottom of the sea (the hydrosphere); and birds and many insects may be compared with the pelagic animals which live in the free waters of the sea. It is an every-day experience that the growth of crops and other terrestrial plants, and the breeding seasons, growth, and migrations of terrestrial animals such as birds, are ultimately dependent on the physical condition of the atmosphere (climate), and yet one fails to realise strongly that corresponding factors in the sea must have similar effects. Temperature, rainfall, and the hygrometrical condition of the atmosphere all affect the life-processes of the animals and plants living on the surface of the land, and apart from actual evidence, one would conclude that temperature and salinity would also affect the habits and metabolism of marine organisms. But in the past marine biology has been studied mainly from the point of view of speciography and morphology, while hydrography has been studied quite apart from biology. It is only within the last two decades that our knowledge of the connection between the life-histories of marine animals and the hydrographical condition of the sea has been acquired, and indeed it is still very meagre.

I will notice very briefly some of the work that has been carried out during recent years, with relation to influence of changes in temperature, salinity, and light, on the metabolism and habits of some marine organisms.

Temperature. It is well known that the distribution and migrations of marine animals are partially dependent on the temperature of the sea. The general facts of the geographical distribution of marine species show this so well that I need only notice the case of the corals. Massive coral formations are only found in those seas where the temperature is not less than 21°C. and where the annual range is not greater than 7°C. The secretion of lime by marine animals is greatly influenced by the temperature of the sea, so that the shells of mollusca are more massive in tropical than in temperate seas. Note the cases, for instance, of the tropical Tridacnas and the mother-of-pearl oysters. Calcareous skeletons are also more attenuated in the case of animals living at the bottom of tropical seas than in the case of those which live at the surface. It is rather difficult to separate the effects of temperature and salinity but there is no doubt that the former is a factor in itself. Thus the changes of salinity in the Irish Sea are insignificant when compared with those in the North Sea and Baltic, but we nevertheless find that there are seasonal migrations of fishes which appear to be related to seasonal changes in temperature. Thus bass (*Labrax lupus*) regularly invade the waters of Morecambe Bay in June and July when the temperature of the sea is approaching the maximum. Soles appear to segregate about the same time, and in the autumn, when the temperature is beginning to fall again, large plaice approach the shore from the deeper waters. Cod come into the Irish Sea when the temperature is falling to the minimum at the end of the year; and whiting frequent the deeper waters in the northern parts of the same sea just after the annual minimum of temperature has been attained, disappearing soon afterwards. D'Arcy Thompson has shewn that there is a certain correspondence in the volume of the catches made by the Aberdeen trawlers, and the temperature of the sea on the fishing grounds frequented by these vessels[1]. Sometimes the abundance of fish on a certain ground appears to be related to the temperature of the water during the preceding season[2]. Many other instances of this kind might be given.

We have seen that the abundance of life in the sea is related to temperature in such a way that the colder seas have a more abundant fauna than the warmer ones. This is well shown in the case of the quantitative plankton catches made in the course of the German "Plankton-Expedition." If the reader will refer to the synoptical chart prepared by Schütt[3] he will see that the larger catches were made in the cold water to the south and east of Greenland, and in the upwelling tongue of cold water which is situated off the Gulf of Guinea; while the poorest catches were made in the warm waters of the Sargasso Sea. These results are apparently paradoxical, for the general effect of increased temperature is the increased metabolism of marine animals. It will be seen, however, that the effect is an indirect one, and depends either on the greater activity of certain forms of bacteria in reducing the amount of available nitrogen food-salts, at higher temperatures; or on the relations between what we may call a general anabolic, as opposed to a general katabolic process in the sea.

Speaking quite generally we may say that the effect of increase of temperature in the sea is to increase the metabolism of cold-blooded animals. The cold winter months are the period of ebb-tide in the life-processes of such creatures. Many fishes live, during the winter, in a state of semi-hibernation, when they are quite inactive, do not feed, do not respire so much as in the summer, and lose weight. When the temperature rises in the spring they become more active, begin to feed, and their metabolism increases. Generally it is during the spring and summer months that reproduction takes place in the sea. In the case of the lower marine plants the enormous reproduction during the

[1] "Aberdeen Trawling Statistics," *Fishery and hydrographical investigations in the North Sea and adjacent waters* (Cd. 2612), 1905, p. 344.
[2] As in the case of the abundance of anchovies in the river Scheldt. *Journ. Mar. Biol. Ass.* vol. I. (N.S.), 1889-90, p. 340.
[3] *Analytische Plankton-Studien*, Kiel and Leipzig, 1892.

spring is to be traced to the accumulation of food-stuffs in the water; no doubt also to the increase in the intensity of sunlight, which raises the intensity of photo-synthesis; and also to the rise in temperature. But the spring spawning habit of the larger marine animals is perhaps the outcome of a long series of seasonal changes in the general external life-conditions of these organisms. Fluctuations in temperature, food supply, &c., have, in the course of time, so operated that a spawning habit has been formed; and heredity has ultimately stamped this seasonal reproductive act on the life-histories of these animals. It has become a periodic habit and is now only indirectly affected by changes in the life-conditions.

There are comparatively few series of observations on the effect of temperature changes on the intensity of marine metabolism, but whatever indicator of the latter process we may take, the intimacy of the connection is well shewn. The pelagic eggs of a teleostean fish hatch out in from one week to three, and the development of the embryo is always hastened by an increase in temperature. So close is the relation between temperature and incubation period that it might almost be expressed mathematically[1]. More precise measurements of the variation of the rate of metabolism with temperature changes are given by estimations of the exchange of oxygen and carbon dioxide between the organism and its medium, and several such series of observations have been made by Pütter. Thus in the case of the metabolism of marine bacteria the following figures were obtained[2]:

Oxygen absorbed by the bacteria contained in one litre of water from Naples Bay, kept in the dark in sterilised vessels.

Temp. Cent.	O used, mgrs.	Temp. Cent.	O used, mgrs.
11·6	0·78	13·3	1·23
12·2	1·17	13·9	1·74
13·2	1·22	14·1	1·58

[1] Reibisch, *Wiss. Meeresunt. Kiel Komm.* Bd. VI. Abth. Kiel, 1902.
[2] *Stoffhaushalt des Meeres*, p. 338.

The results of this series of estimations are of great interest, for they indicate that the bacteria may exert a powerful influence in reducing the proportion of oxygen contained in the warmer seas. The mean value as indicated by the above figures at 13°·1 C. is 1·2 milligrammes, and from this Pütter calculates that 1000 millions of bacteria—a number which may be present in one cubic metre of Mediterranean water—may use up about 300-times their own weight of the dissolved oxygen.

So also with the other organisms dealt with by Pütter. Thus the calcareous alga *Lithothamnion racemus*, when kept in the dark, uses up oxygen, and the amount of the latter utilised by the plant is almost exactly proportional to the temperature, increasing with the latter. The oxygen-consumption of the holothurian *Cucumaria grubei* increases with the temperature, and as we have seen (p. 232) the nature of the nutrition process also changes, so that at the higher temperatures the oxidation processes become more prominent than those of the fermentation character[1].

Salinity. Precisely how changes in the concentration of the salts of the sea water affect the habits and metabolism of marine animals we do not know. It is of course easy to see how changes in the density of the water may have powerful effects in the distribution of certain animals in the sea. Many organisms, such as pelagic fish eggs, have almost exactly the same specific gravity as the water in which they float; and any alterations in the density of the latter (such as may be caused by changes in salinity, in temperature, or both) may cause these organisms to rise or sink in the sea; and since the strength and direction of the currents may vary at different levels, the distribution of the fish eggs, or of the larvae hatched from them, may be affected. Further, such eggs or larvae may be carried by currents into other parts of the sea, the estuaries of rivers for instance, where the density is so low that they at once sink to the bottom, become smothered in mud, and die. It is probable that considerable numbers of pelagic fish eggs may be destroyed in this way in the Baltic, where the density of the water may fall so low that fish eggs and larvae do

[1] *Vergleich. Phys. Stoffwechsel*, p. 52.

not float in it. These are indirect effects of changes in the salinity of the sea water; but there is also evidence that slight changes in the salinity *per se* may affect the metabolism of marine organisms. How precisely this happens we do not know, but we do know that very slight changes in the salinity and composition of sea water, in which float the eggs or larvae of certain marine animals, affect, in a very notable manner, the further development of these things[1]. *A priori* we should expect that reactions of this nature would take place, for in many marine invertebrates the concentration of the juices of the body is very much the same as that of the salts in sea water. In these creatures the wall of the body is a "semi-permeable membrane," and diffusion out of, and into the blood, modified of course by "vital action," must take place. Probably the key to many of the effects which we have to consider may be sought in these diffusion phenomena.

However this may be there is good evidence that slight changes of salinity have considerable influence in determining the movements of marine animals. So far this evidence relates almost entirely to the migrations of the edible fishes, and has been collected by a comparison of the fishery statistics with the results of the hydrographic study of the northern seas.

(1) The migration of the herring. This is one of the older problems of the natural history books. The great summer herring fishery of the coasts of Britain begins about May off the coasts of the outer Hebrides and Orkneys, extends, in June, to the sea off the coasts of Aberdeenshire; and then changes in such a manner that the bulk of the fish taken in each successive month are caught further to the south along the east coast of Britain, until in November the fishery is concentrated in the southern North Sea, and in December and January off the coasts of Devon and Cornwall. Nothing can be clearer than this orderly change in the situation of the place of the maximum herring fishery. It used to be supposed that there were great shoals of herring which came from

[1] The reader will remember the classical experiments of Loeb, who induced the eggs of the sea-urchin to begin to develop merely by adding certain salts, in slight traces, to the water in which these eggs were contained. See also some very interesting observations by Moore, Roaf and Whitley on the influence of slight traces of acids, alkalies, &c. on the development of the eggs of the plaice and sea-urchin. *Proc. Roy. Soc. London*, January 6, 1906.

somewhere in the Arctic regions, and that as the season progressed these fishes gradually migrated to the south. It is now fairly certain, however, that the herring which appear off the east coast of Britain are mostly local in their origin; and that they form shoals at certain times in the year for the purpose of breeding. It is only when these shoaling movements are taking place that the herring are sufficiently crowded together to allow of the practice of the methods of fishing. But it is now nearly certain that the shoaling migrations are to be associated with changes in the salinity and temperature of the sea water. If the reader will refer to a chart of the positions of the great summer herring fisheries[1], and then compare this with the chart of the temperature and salinity of the sea during each successive month, he will just be able to see that as the isotherms 13° C. to 15° C. pass to the south, that is as the Atlantic water of the European stream passes to the south of the North Sea from the north, so also does the position of the maximum herring fishery change. But the correspondence is not an exact one, and one must not conclude, on this account, that there is no connection between the temperature and salinity of the sea and the shoaling of the herring. It is far more probable that the lack of exact correspondence is due to the defective fishery statistics, on the one hand, and the infrequency of the hydrographic observations, on the other.

(2) Winter herring in the Skagerak. The winter herring is a fish which does not inhabit the Atlantic water proper (salinity 35 per 1000 and over) but a mixture of this with coast water, which is called "bank-water," and which has a salinity of about 32 to 33 per 1000. When this layer of bank-water is abundant in a sea-area then the winter herring is also abundant; and when the bank-water is scanty, or absent, then the winter herring is also rare, or absent altogether. Such a scarcity of bank-water may be due to the exceptional flooding of the sea-area with the Atlantic water from surface to bottom, and when this occurs the winter herring fishery is a failure[2].

[1] See Kyle, *Bulletin Statistique, Cons. Perm. Internat. Explor. Mer*, vol. I. 1906, where there are charts of the situation of the herring fishery, from month to month, during the year. The hydrographical data may be found in the *Bulletins des Resultats*, also published by the International Council for Fishery Investigations.

[2] Pettersson, *Rappts. et Proc.-Verbaux*, vol. III. Appdx. A, 1905.

(3) Line fisheries in the North Sea. If the reader will refer to the chart of depths of the North Atlantic he will see that north from Shetland the bottom of the North Sea slopes down gradually into the depths of the Norwegian Sea. During most of the year Atlantic water lies on this slope of the North Sea Plateau; at the bottom there is cold Arctic water; while between this, and the surface stratum of Atlantic water, there is a mass of water which is intermediate in salinity to the two principal layers. This stratum of mixed water is rich in copepods and other plankton, and is inhabited by fishes such as the halibut and ling. Sometimes the upper salt layer is thick and then the mixed layer lies far down the slope; sometimes it is thin, and then the intermediate layer lies further up the slope. In June, when the Atlantic flooding is at a minimum, the 35 per 1000 layer is thin and the fishermen set their lines in comparatively shallow water. In the autumn it is thick, and the lines are set in deeper water[1].

(4) Fisheries of the Barentz Sea. Every year the Barentz Sea is flooded with relatively warm and salt water from the European Stream, and every year a fishery for food fishes, such as plaice, recurs in these remote seas. When the Atlantic water sets in there is an immigration of food fishes into the Barentz Sea. On the subsidence of this flooding of genial water the food fishes desert the area, the trawling is a failure, and the benthic fauna changes in character[2].

(5) Icelandic fisheries. Iceland is surrounded by water which is part of the southerly-flowing polar stream. On its seaward border this cold coastal water is mixed with water which is warmer and denser, and which is part of the Irminger Current, an offshoot of the European Stream. Just on the border region of the two areas of water there is a mixed portion, where the salinity of the Atlantic constituent is reduced slightly by the fresher Polar water. Here the spawning cod is to be found, and this fish does not inhabit the colder inshore zone. "In the course of the summer all food fishes inhabiting this border region (of mixed water) pass to the east with the general movement of the stream[3]."

[1] Pettersson, *Rappts. et Proc.-Verbaux*, vol. III. Appdx. A, 1905.
[2] Breitfuss, *Verhandl. Vth Internat. Zool. Congress.*
[3] Schmidt, "Fiskeriundersogelser Island Faeroerne Sommer 1903," *Skriften Komm. Havundersogelser*, I. Kjobenhavn, 1904.

(6) "North" and "South" fishes. Long ago Heincke and Möbius[1] divided the fishes of the Baltic into "north" and "south" fishes. More lately Henking made a somewhat similar classification. The terms are relative ones, and apply only to the general habits of the fishes. South-fishes, or summer-fishes, are those which inhabit the water of the European Stream and its offshoots, or live in proximity to this water mass. The best examples of these fishes are the garfish (*Belone*), the hake, the sole, the turbot, and the summer herring. North-fishes, or winter-fishes, inhabit the mixture of the Atlantic and Arctic waters. The best examples are the cod, the ling, the halibut, and the winter herring. Fishery statistics shew that the largest catches of these fishes are made in the North Sea, at the times and places when the water of each type predominates[2].

Sunlight. I have already indicated (in Chap. VII.) the manner in which various marine organisms react towards changes in the intensity of the light falling upon them. Such reactions may be called "heliotropic," or "phototropic," and are really reflexes towards changes in external stimuli, and may be compared with the nervous process which impels a moth to dash into a naked flame, or a sea-bird to fly against the windows of a lighthouse. The crowding of the trochospheres of *Phyllodoce* against the best illuminated side of the vessel of water which contains them; the similar behaviour of copepods, which has been observed both in aquaria and in the sea; and the distribution of *Sagitta* with respect to the depth of water, or the amount of sunlight, are probably instances of nervous reflexes. They are due to the stimulus of the peripheral nervous system, and the conversion, in the central nervous system, of the impulses thus generated into motor impulses, which cause the organism to move towards, or away from the source of light.

It is quite different with regard to the behaviour of the marine protophyta towards changes in the intensity of the incident light. Perhaps one ought not to say that these organisms are not affected by light in the same manner as that in which a copepod is affected,

---

[1] *Bericht Untersuchung. deutschen Meeres*, II. p. 278, Kiel, 1877–81.
[2] Henking, *Rappts. Proc.-Verb.* vol. III. 1905.

CH. X]   THE CONDITIONS OF LIFE IN THE SEA   249

for their protoplasm possesses irritability, and so may possibly react towards light stimuli. But the main influence of light on the metabolism of the diatoms, and other protophyta, is expressed in the power of photo-synthesis which these organisms possess. This influence of light is manifested in the elaboration of carbohydrate from the carbon dioxide and water of the medium, and in the building-up of proteid from these substances and from the inorganic food-salts of the sea. We shall see that the tendency of all metabolic processes in the sea is towards the resolution of the tissue substances of organisms into the most highly organised compounds of nitrogen and carbon—that is into nitric acid, water, and carbon dioxide. These substances can only be utilised by organisms when the energy of sunlight is intercepted and used in the process of photo-synthesis by the green plant. Thus all the energy manifested in the organic world is ultimately derived from the energy of solar radiation.

What precisely are the relations of the metabolism of marine protophyta towards the intensity of sunlight, or radiant energy of different wave-length, have not been sufficiently investigated. Much work of this kind has been done by the plant physiologists, in the case of terrestrial plants, but we cannot assume that the results so obtained hold good with regard to the unicellular plants living in the sea; and this caution is all the more necessary when we remember that the experiments of Pütter on the metabolism of marine organisms have led us to suspect that the life-processes which are characteristic of the higher land animals do not necessarily hold good with regard to the lower animals living in the sea. We have seen that the Plankton-Expedition found *Halosphaera* living at a depth of over 1000 metres, and this would apparently indicate that the amount of light necessary for the carrying on of photo-synthetical processes may be very small; for at this depth there must be almost total darkness, unless the light of the phosphorescent organisms is sufficient for the needs of the plants. Pantanelli[1] has shewn that there is an "optimum" intensity of light in the case of the photo-synthesis of starch by the land plants, which is about one-quarter of the intensity of full sunlight; and it is not improbable that there may

[1] See Jost's *Plant Physiology*, Clarendon Press, 1907.

also be an optimum intensity of light in the case of the marine protophyta. It may sometimes be observed that a culture of marine diatoms in sea water mud exposed to strong sunlight does not shew the same obvious indications of rapid multiplication as when the same culture is shaded from the direct rays of the sun.

Neither must we assume that photo-chemical reactions—the power of making use of solar radiations—are exhibited only by marine plants, or at least by those organisms possessing chlorophyll plastids. Patten[1] suggested long ago that the "eyes" of the scallop (*Pecten*) were not solely visual organs, but also constituted an apparatus for the utilisation of the energy of sunlight; and there is a fair amount of experimental evidence in favour of this hypothesis[2]. It must have occurred to many zoologists that if a principle of "economy of effort" is to be assumed in morphological speculations, the explanation of the functions of a battery of highly developed visual organs in such an animal as *Pecten* is rather difficult to find. But however this may be we have direct evidence that the intensity of metabolism in some marine animals may be influenced by the amount of light which falls on their surface. I quote here some results obtained by Pütter[3] in the course of a research on the amount of oxygen used by the sponge *Suberites* in conditions of light and darkness:

Amount of oxygen used per hour, and per kilogramme of weight.

Temperature, Cent.	Dark, O used up, mgrs.	Light, O used up, mgrs.
17·7 to 17·9	15	41
12·8 to 13	68	131
12·3 to 12·5	43	57

There can apparently be no doubt of the existence of a photo-chemical activity in the metabolic processes of *Suberites*. The adult sponge is an organism possessing no powers of locomotion, so

[1] *Mitth. Zool. Stat. Neapel*, Bd. VI. 1886.
[2] See Herbst, "Über die Bedeutung d. Pigments f. d. physiol. Wirkung der Lichtstrahlen." *Zeitschr. Allgem. Phys.* vol. VI. 1907, p. 44.
[3] *Vergleich. Phys. Stoffwechsel*, p. 30.

that there can be no question of an increase of energy due to the stimulation of a sensory surface by light, and of a consequent greater amount of movement producing an increased respiratory interchange. Further there is no differentiated sensory, or central nervous system. We must apparently conclude that the change in the intensity of light produces a change in the rate of metabolism, and that the increased oxygen consumption is the indicator of these changes.

There is a relation between the colour and transparency of sea water and the density of the plankton. Sea water varies in colour from cobalt-blue to yellow, and these variations are due to the mixture of the blue colour, which is due to the water itself, with the green-yellow to brown-yellow of the chromatophores of the plant organisms. The transparency also varies, as we have seen, and the changes are due to the variable amount of inorganic particles in suspension, and to the varying opacity of the organisms of the plankton. There is a parallelism between the density of planktonic life, the transparency, and the colour. "The yellow Baltic only allows us to see the white plankton-net at a few metres beneath the surface; the green Arctic water transmits light rays from a greater depth; and what visitor to the Mediterranean has not been impressed by the blue colour of that pellucid sea?" The Baltic shews us a "colossal wealth" of plankton life; the Arctic waters, as in the Irminger Sea, or the fjords of Greenland, are not much poorer in vegetable microscopic organisms; the plankton fauna and flora of the Mediterranean is still more meagre; and that of the oceanic Sargasso Sea is apparently the poorest of all. "Observations of transparency, colour, and plankton-contents, all shew a parallelism, and all lead to the same conclusion, that the pure blue is the colour of desolation of the high seas. The colour of the Arctic seas is comparable with the green of the meadows, but the colour of the most luxuriant plankton-flora is the yellow of the Baltic[1]."

We have considered only the coarser factors of temperature, salinity and light, but it would probably be wrong to conclude that

---

[1] Schütt, *Das Pflanzenleben der Hochsee*, p. 76, Kiel u. Leipzig, 1893.

only these influence the metabolism of marine organisms. We know that meteorological changes, so subtle as almost to defy analysis, do affect man and the higher terrestrial animals. Every angler knows that the "rising" of trout is affected by almost imperceptible changes in the condition of the atmosphere. And since the physical organisation of many of the marine animals is hardly, if at all, inferior to that of the terrestrial mammals it may be the case that very slight changes in the physical condition of the sea water affect them also.

CHAPTER XI.

BACTERIA IN THE SEA.

So far I have alluded only very casually to the bacteria of the sea. Because of their universal distribution and their special modes of nutrition these micro-organisms do not fit into any general scheme of classification such as we have considered with regard to the other organisms in the sea. They belong to the benthos, for we find them in the muds and oozes at the sea bottom; but they belong also to the plankton, for we find them everywhere diffused throughout the water of the sea. Both in the sea and on the land the bacteria are quite ubiquitous, and there is hardly any object in animate and inanimate nature which does not harbour them. They abound on the surfaces of both terrestrial plants and animals, and they are found in the cavities of the latter which are in communication with the outside world; while in some diseased states almost every organ and tissue of the animal body may contain them. It is because of their universal occurrence in nature, and also because of their peculiar reactions toward dead organic substance and toward the inorganic food materials of organisms, that the bacteria are of such extraordinary importance in the "household of nature."

Note that quite special methods of investigation have to be adopted in studying them. No net that can be devised is fine enough to catch marine bacteria. We can indeed separate them from water by means of porcelain filters, but this is hardly a means of collecting them. They are always obtained from sea water, or from mud, or from the surfaces or parts of the bodies of organisms, by taking a small portion of any of these materials and implanting this in a solution containing certain food materials which can be utilised by the bacteria. Multiplication of the organisms there takes place and their numbers become so enormously augmented that their study is facilitated. Or if we wish to estimate the numbers in which they occur in water or in any other material a sample volume or weight of the latter is taken and is distributed throughout the substance of a plate of some solid food medium, such as nutrient gelatine or agar, or is spread on the surface of the latter. Multiplication then takes place as before; each separate bacterium gives rise to a colony which is composed of a vast multitude of individuals; and since the colonies are large enough to be easily visible to the naked eye the number of individual bacteria in the original sample can be deduced, for each bacterium has given rise to one colony. Usually the colonies consist of individuals of the same species, so that the different species present in the crude sample can be isolated from each other and studied separately. The nature of the food media on which bacteria grow and multiply; the temperature at which they grow best, or are killed; the appearance of the colonies; the reactions of the germs to various chemical substances; and the microscopic characters of the organisms; all differ with the species, and are made use of in determining the latter[1].

The bacteria may be classified according to their morphology but this only affords a general scheme of grouping. Fischer distinguishes the following families:

(1) The *Coccaceae*, in which the germs are usually spherical bodies of minute dimensions. The individuals may be separate or arranged in little clusters, as in the case of the *Staphylococci* of the pathologists. In addition to these micrococci we have the *Sarcinae* and their allies, in which the cocci are arranged in groups of four, and the *Streptococci*, in which they are arranged in chains. These characters depend of course on the direction of the

[1] It is of course quite impossible to allude to the methods of the science here. During the last forty years the methods of bacteriology have been dominated by the immense importance which the study of micro-organisms has acquired from the standpoint of medical science. The reader who is desirous of studying bacteria from the point of view of the circulation of matter in nature cannot do better than study A. Fischer's *Structure and Functions of Bacteria* (English translation by Coppen Jones, Clarendon Press, 1900).

divisions in which the original bacterial cell, which gave rise to the colony, cleaved. (2) The *Bacillaceae*; these are the typical bacteria: they are rod-shaped bodies, usually cylindrical, but sometimes egg-shaped and difficult to distinguish from cocci. (3) The *Spirallaceae*: in this family the vegetative cells are always cylindrical, but they are spirally twisted rods, or bent, or slightly twisted "commas," or long and slender spirals or Spirochaetes. Some of these latter organisms belong to other groups of protista. Finally there are the (4) *Trichobacteriaceae*, among which the vegetative form is a long branched or unbranched filament, sometimes rigid and enclosed in a sheath, and sometimes motile with undulating or gliding movements. Individual cells break off from these filaments and each of these, the swarm-spores, or gonidia, gives rise to a new colony. They are the most highly organised of the bacteria. In Fig. 31 I give an illustration of some of the typical forms of bacteria. The reader must note that these morphological characters carry us only a little way in the determination of the species to which a bacterium belongs. The biology of the organisms; their effect on different food media; the nature of their sporulating process; their reaction towards oxygen, temperature, &c.; their pathogenic effects: these and many other properties all are made use of in the identification of the species. Most bacteria are motile, that is they can execute movements, and these are caused by cilia or flagella. Many, or most, produce spores, which consist of the protoplasm of the cells concentrated to form small rounded masses possessing a thick wall or capsule. They constitute the resting stage of the organism and they are always highly resistant to agencies which destroy the bacterial vegetative cell. Some bacteria—anaerobes—can live in the total absence of oxygen, while others live preferably in the presence of this gas, and may become attenuated or enfeebled if the amount of oxygen in their atmosphere diminishes—between aerobic and anaerobic bacteria there are many gradations. For every bacterium there is an "optimum" temperature at which the organism grows best; a maximum temperature at which it is killed; and a minimum temperature at which it may survive, but at which it cannot grow. Usually the multiplication of a bacterium is arrested at a temperature of $0°C.$; and at a temperature above $42°C.$

many are killed, but some can survive a temperature of about twice that mentioned. The spores of bacteria can resist freezing, drying, or temperatures which destroy the vegetative cells: thus the spores of anthrax bacteria can withstand immersion in boiling water for some time. A very great number of observations

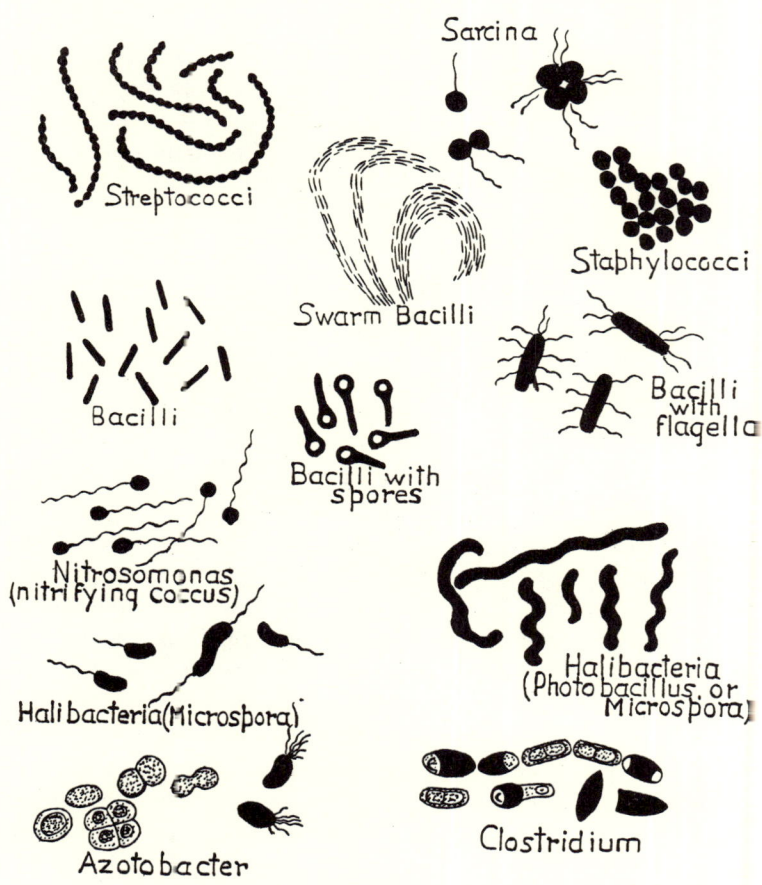

Fig. 31. Forms of bacteria. All are highly magnified, but the scale varies.

have been accumulated with regard to the behaviour of bacteria towards various nutrient materials, towards chemical reagents, towards the fluids of the body in normal and diseased states, and with regard to their disease-producing effects; and pathologists

The distribution of bacteria in the sea. There are many isolated observations of the occurrence of bacteria in the sea, but the first really extensive investigation of their presence was made during the Plankton Expedition[1]. Generally speaking the water near the shore is richer in bacteria than that further out at sea, and much richer than the water of truly oceanic areas. This abundance of germs near the land is due to the drainage from the land; and also to the rich algal flora of the shore which provides abundant food for the micro-organisms. Estuaries on densely populated parts of the coasts are rich in those bacteria which abound in the excreta of man and the domestic animals. I have estimated as many as 70 bacteria belonging to the *Bacillus coli* group in 1 c.c. of the water of such an estuary. But this influence of the land does not extend to more than about ten miles out to sea. Thus I have made several analyses of the contents of the bodies of shellfish living at this distance from the shore and have failed to find any evidence of the presence of intestinal bacteria, although the shellfish near the shore on such a coast are universally infected with such bacteria. Generally speaking it is the case that the bacteria which inhabit the intestines of land animals do not find a suitable habitat in sea water and when introduced into the latter they gradually die out. The experiments made at the instigation of the Royal Commission on Sewage Disposal shew that the pathogenic and non-pathogenic bacteria of man and the higher animals do not multiply in the sea—a very fortunate thing in these days of crowded watering places, and popular shellfish consumption.

Such bacteria do not occur at all in the open sea. The forms which we find there are quite different. Sarcinae and Streptococci were not found at all by Fischer in the open Atlantic, and other cocciform bacteria occurred very sparingly. Moulds like *Penicillium* and *Aspergillus* were observed, but it is possible that the occurrence of these was due to the accidental contamination of the water samples and cultures. *Saccharomyces*, or yeasts, were found

[1] B. Fischer, *Ergebnisse Plankton Expedition*, Bd. IV. 1894.

at great distances from the land. Bacilli occurred but they were not numerous. The predominant bacterial species in the open sea are spirallar or bent rods, chains of commas, or twisted corkscrew-like forms. Fischer shews that there is great diversity in the morphological characters of oceanic bacteria, but that, as a class, they are distinguished by the following features: (1) they are either bent rods or spirals; (2) they exhibit a decided preference for food solutions containing salt in the proportion in which this occurs in sea water; and (3) they are often self-luminous. It is true that many land forms exhibit these characters: thus the cholera germ, and the associated group of vibrios, have the form which is characteristic of the oceanic species; some land forms can live in highly saline solutions; and Kutscher has shewn that a bacillus may be present in the human intestine that may be self-luminous[1]. But as a group these characters distinguish the truly oceanic from the land bacteria.

Halibacteria. This is the general term (quite a loose one of course) which has been applied to the marine bacteria. They can live and multiply in sea water containing no added food salts or other substances. This was repeatedly proved on the Plankton Expedition when water taken from the sea was at once placed in sterilised test tubes, and in a few days it was found that decided multiplication had taken place. They are almost exclusively aerobic bacteria, living in the presence of oxygen, but many can function as facultative anaerobes, that is they can live in solutions containing no free oxygen. Their reaction to temperature varies greatly, thus *Halibacterium pellucidum* has a wide range of temperature, some luminous bacteria grew and produced light at the freezing point of fresh water, others multiplied at 3° to 5° C., others again at 10°, and one even at 46° C. In no case did Fischer observe endogenous spore formation, that is the formation of a spore within the vegetative growing cell; nevertheless he was able to keep some of his cultures on agar for 2½ years. All halibacteria are more or less motile. Some appear to be pathogenic: when introduced into the peritoneal cavity of guinea-pigs or mice, death of the host took place and the bacterium could be obtained from

[1] *Deutsche medicinische Wochenschrift*, No. 49, 1893.

the heart's blood of the animal shewing that it had multiplied in the tissues.

The distribution of the halibacteria is very wide. *H. polymorphum* is the most ubiquitous form and then comes *H. roseum* and *H. pellucidum*. They occur at all levels of the sea down to 1000 metres. But whereas the bacteria taken from the surface of the sea did not multiply in pure sea water, those obtained from the depths did so. Possibly the strong light at the surface enfeebled them, for at the depths from which the more healthy individuals came there could be little or no sunlight. They occurred on the exterior of some fishes and other marine animals taken during the Plankton Expedition.

Luminous bacteria. Many marine bacteria produce light just as do hosts of other organisms like *Ceratium*, *Noctiluca*, worms, deep sea fishes, crustacea, and other creatures. In the North Sea and Baltic phosphorescence is usually due to *Ceratium*, and in the Irish Sea to *Noctiluca*. Often however the light is due to luminous bacteria (*Photobacillus* or *Microspora*). Often fish such as herring, mackerel, whiting, &c. are brilliantly phosphorescent, even before decomposition is very evident. I have seen the liquid (water containing alcohol from which most of the latter had evaporated) in a tank containing very imperfectly preserved whiting so very phosphorescent that it emitted quite a strong light. This phenomenon is due to the enormous multiplication of organisms which have been called Photobacteria, the commoner species being *P. indicans, P. luminosum, P. phosphorescens*, &c. In a brilliantly phosphorescent sea the light is only shewn on the surface where the waves break in contact with the air, or are churned up in some way, as by the blade of an oar, or agitated as in the waves thrown off from the bows of a ship, or in the wake of the propeller or paddles. In these cases the luminosity is evoked by the supply of oxygen to the bacteria. But one can often see a net towed beneath the surface of the sea brilliantly phosphorescent. In some cases the phenomenon is certainly due, not to bacteria, but to *Ceratium* or *Noctiluca*. But whether it is occasioned by bacteria or protozoa the cause is always an oxidation process, and in the case of all organisms displaying it the light is produced as part of the surplus energy of the organism. In the case of the bacteria it is arrested by withdrawal of the oxygen supply, and is increased by the abundant supply of the latter, and also by the supply of some highly combustible food such as carbohydrate. In all cases the light is greenish blue in colour.

The nearest terrestrial allies of the halibacteria are the cholera bacillus and its congeners. In these two groups of micro-organisms the appearance of the colonies to the naked eye is similar, and the morphology and biology are the same, or are very similar. The cholera germ is a typical fresh water microbe. Fischer points out that the spirallar form is the best adapted for progression through water or for suspension in the latter.

In what numbers are these bacteria present in the sea? Fischer estimated that from 20 to 785 germs were present per c.c. of the water from the Sargasso Sea in the mid Atlantic. In the Gulf Stream 345 bacteria were counted in 1 c.c. on one occasion. If we take it that about 600 germs are in 1 c.c. then we have a total of 600 millions per cubic metre of sea. I do not doubt that this is a reasonable estimate, and that it is vastly exceeded in the water from inshore areas. But prodigious as are the numbers, they represent only a very small mass of life. Thus 600 millions of bacteria form far less than one gram of solid substance.

The putrefactive bacteria. We may now consider some groups of marine bacteria which have special modes of life, and the importance of which lies in the manner in which these organisms affect the abundance of the ultimate sources of food stuffs in the sea and on the land. Those germs which we have been considering so far are concerned with the breaking down of dead organic matter, that is we call them putrefactive bacteria. This leads us to a more minute consideration of the modes of nutrition of these micro-organisms. Fischer divides all the bacteria into three main divisions according to their habitat and mode of nutrition. These are:

(1) *The prototrophic bacteria.* These are the most lowly organised of living things. They can use as sources of food such very simple substances as nitrates and nitrites, carbon dioxide, sulphuretted hydrogen and even elementary nitrogen. Green plants

can also use as food materials the nitrates and nitrites and carbon dioxide, but the activity of the plant, so far as its carbon assimilation is concerned, is bound up with the possession of chlorophyll, which substance enables it to utilise the energy of sunlight. Now the prototrophic bacteria have no chlorophyll; they can live and grow in the dark; and they can synthesise the complex proteid substance of their living bodies from nitrates, nitrites, carbon dioxide and a few simple mineral salts. No other organisms can do this. They live in the open and are never parasitic.

(2) *The metatrophic bacteria.* Most bacteria belong to this division. They live upon the dead organic substance of plants and animals. They can form enzymes or ferments which bring about important chemical changes in many substances. They have many sources of food, but this is always highly complex compared with the food-stuffs of the prototrophic germs. They may live as parasites within the animal body or on the surfaces of the bodies of both plants and animals. They are sometimes parasites but they can also live in the open.

(3) *The paratrophic bacteria.* These also live upon the dead or diseased organic matter of the plant and animal organism, and they are always parasites inhabiting the tissues or cavities of the living body. Like the metatrophic bacteria they demand as food-stuffs the highly complex substances of organised matter. They never live in the open. Most disease-producing bacteria belong to this division.

The putrefactive bacteria belong to the metatrophic group. When the body of a plant or animal dies it may, or it may not, decompose. If the dead body is kept at a low temperature (as in the case of the famous Siberian mammoth) it will not putrefy. If it is kept at a moderately high temperature (say 50° C.) it will not putrefy. If again it is permeated with certain substances, such as corrosive sublimate, or carbolic acid (antiseptics), neither will it putrefy. Decomposition is always due to the living activity of metatrophic bacteria, and in the above circumstances the bacteria are either destroyed, or their activity is inhibited or is arrested.

When the proteid substance of the body is decomposed by boiling acid, or caustic potash, or barium hydrate, certain substances result from the break-down of the proteid molecules. These are very numerous and are still very complex chemical compounds. A number of substances called amino-acids are produced, some of which are leucine, tyrosine, glycine, &c. Albumoses and peptones, very complex substances, may also be produced during the first stages of these decompositions. Very malodorous substances such as indol and skatol, substances to which the odour of the excreta is due, are also produced. It is only during the last few years that we have become intimately acquainted with the manner in which the frightfully complicated proteid molecule-complex decomposes, and the wonderful researches of Emil Fischer, Abderhalden and others indicate that the actual manufacture of proteid substance in the laboratory, with all the results that this achievement suggests, may be a thing of the near future[1].

Now when putrefaction begins in a dead body much the same series of compounds as are produced by the action of chemicals (in the hydrolysis of the proteid) come into existence. The bacteria seize on the dead substance and break it down. Some of the putrefactive bacteria—the *saprogenic* species—break down the proteid substance into simpler stuffs, and others—the *saprophile* species—utilise these first products of decomposition as food. Both processes go on together. Putrefaction is a very complex process and it is modified by many agencies such as the temperature and the presence or absence of oxygen. When oxygen is freely admitted to the putrefying mass—aerobic putrefaction—rapid decomposition with but little smell takes place. But when air is partially or entirely excluded—anaerobic putrefaction—as in the interior of a decomposing carcase, then foul-smelling products accumulate. Sometimes ptomaines, many of them highly poisonous substances, are formed. Altogether the putrefactive process is very complex, and our knowledge of it is still very imperfect.

But in the long run all putrefactive processes due to bacteria have the same result. The complex proteid substance is broken down into a few comparatively simple substances. These are ammonia, carbon dioxide, sulphuretted hydrogen, water, phosphoretted hydrogen (which at once becomes oxidised by the oxygen

---

[1] An excellent account of these recent investigations is given by Plimmer in *Science Progress* for July 1907.

of the air), marsh gas, nitrogen and possibly free hydrogen. The sulphuretted hydrogen becomes oxidised in various ways, and then all that remains of the proteid are a few innocuous and inoffensive compounds, which rapidly diffuse into the water of the soil, into rivers, &c.

Fermentation bacteria. In the organic body there are three great classes of compounds, the nitrogenous proteids, the non-nitrogenous carbohydrates, and the fats. The proteids are the "flesh" of the body, and many other tissues not popularly termed "flesh," such as horn, chitin, &c. The carbohydrates are starch, sugar, cellulose, &c. The fats are compounds of a fatty acid with glycerine. Putrefactive decomposition concerns itself with the proteids. But the carbohydrates and the fats are also broken down by the agency of micro-organisms. The fats are split into their fatty acids and glycerine, and then each of these compounds is broken down by a process which we distinguish from putrefaction and call fermentation. Fermentative bacteria doubtless exist in the sea though we have little knowledge of their distribution.

Fermentation processes are among the most common things of everyday life. The formation of alcohol in the manufacture of wines, spirits and beer, is a process of fermentation. The souring of wine or beer, or of milk, is also a fermentation. So is the production of vinegar, the rising of bread in baking, the ripening of cheese, and many other processes used in the domestic arts. In all fermentation processes some carbohydrate substance is decomposed by the vital action of bacteria and a simpler substance or many simpler substances are formed. Always, or nearly always, carbon dioxide is also formed. But whereas the end products of a putrefactive process are ammonia and sulphuretted hydrogen and some other bodies, those of fermentation are carbon dioxide and water.

The characteristic feature of a putrefactive or fermentative process is that a large amount of the food material of the micro-organism (bacterium, mould or yeast) is broken down in order to obtain a comparatively small amount of energy. In the assimilation of sugar, as a food, by a higher organism this substance is completely oxidised and the products of the combustion appear as carbon dioxide and water. All the available energy of the food substance is thus obtained by the organism. But in the case of the fermentation of sugar by the wine yeast this complete utilisation of the sugar by the organism does not take place, for a certain portion of the former (say one half) is oxidised as far as alcohol, and about one half suffers complete oxidation to carbon dioxide. Many minor products are of course obtained in small quantities. So also with putrefaction. A great number of molecules are oxidised, or otherwise broken down, and in these incomplete decompositions the micro-organism obtains its energy by, in one sense, a wasteful reaction. Many different bacteria and other organisms are concerned in putrefaction and fermentation, and the one may begin where the other leaves off. So in the end the products are the completely resolved ones, ammonia, carbon dioxide, water, &c.

The sulphur bacteria. In some parts of the sea, as for instance in the "dead grounds" of the Bay of Kiel, in some parts of the Black Sea, and perhaps in parts of some of the Norwegian fjords, where the water circulation is defective, and where there may be a deficiency of oxygen, very remarkable bacteria are to be found. These are the sulphur bacteria, the occurrence of which is not, however, confined to these habitats. In the places I have mentioned sulphuretted hydrogen is evolved from the decomposition of dead organic matter, and this sulphuretted hydrogen, to us a vilely smelling and poisonous gas, is utilised as food substance by the bacteria. Such a microbe as *Beggiatoa* takes in the SH_2 and oxidises it so that the sulphur is deposited in the cells of the bacterial colony, and the hydrogen appears as water. This is the form of assimilation of the organisms. Then some of the sulphur thus resulting from the decomposition of the SH_2 is oxidised to sulphuric acid. This is the form of respiration of the organism. It requires some source of nitrogen for the formation of its living proteid and this it obtains from the minute quantities of nitrates and nitrites which exist in solution in the water in which it lives. But it requires very little nitrogen compound, for whereas a higher animal may require to oxidise some of the living nitrogenous tissue of its own body in order to obtain its energy, the sulphur bacterium oxidises the sulphur stored in its cells as the result of the assimilation of the SH_2. Thus the proteid part of the cell is protected from waste, and the minimal quantity of nitrogenous food-stuff suffices.

This process is exactly reversed by the desulphurising bacteria, of which one species—*Spirillum desulphuricans*—is known. This organism is able to break down sulphates in its cells, liberating the sulphur in the form of sulphuretted hydrogen.

The nitrifying bacteria. Thus the dead matter of organisms is broken down by the vital activity of the bacteria. In the case of putrefactive decomposition the process is carried out in a number of stages, but the final result is the conversion of the complex proteid substance into ammonia. In the case of the fermentation process the fats and carbohydrates are so broken down that the elements of these substances finally appear as carbon dioxide and water. But this is not all. The ammonia of the putrefactive process is further transformed and is oxidised by the vital activity of the nitrifying bacteria.

We know most of the details of this process now but it is only as the result of one of the very finest pieces of biological work that have ever been carried out. Winogradsky's work has now become classical and perhaps some day his investigations on nitrification will be translated into English and presented to the student of biology as a model of an investigation and an ideal[1]. Previously it had been assumed that the accumulations of nitrogen salts—the nitre beds of Chili, for instance—had been formed by the oxidation of dead organic matter, by the oxygen of the atmosphere. Winogradsky, after a long series of investigations, in the course of which fundamental methods of investigation had to be evolved, shewed that the transformation of the organic matter was the result of bacterial activity.

Only three genera of nitrifying bacteria are known but they are of world-wide distribution. They are *Nitrosomonas*, *Nitrococcus*, and *Nitrobacter*. *Nitrosomonas*, which (see Fig. 31) is a small oval bacillus furnished with a long cilium, oxidises ammonia to nitrous acid. *Nitrobacter* is a minute, non-motile bacillus and its function is to oxidise nitrous to nitric acid. These two

[1] There is a good summary of Winogradsky's investigations on nitrifying bacteria in Jost's *Plant Physiology*.

bacteria are, with regard to their mode of nutrition, and probably also with regard to their morphology, absolutely the simplest organisms known to us. They have no chlorophyll and can live and multiply absolutely in the dark. They require as food only ammonia or nitrous acid as nitrogen sources, and carbon dioxide as a source of carbon. With these extremely simple substances and with traces of the essential mineral salts they can form proteid substance and convert this into living protoplasm. They are aerobic germs deriving their oxygen from the atmosphere. Their carbon dioxide is also obtained from the air or from the ground water, and from the latter source they also obtain their mineral salts. Their energy is derived from the oxidation of the ammonia or nitrous acid.

Nitrification of the ammonia resulting from the putrefactive decomposition of dead proteid is carried out in two stages: (1) the oxidation of the ammonia to nitrous acid, and (2) the oxidation of the latter compound to nitric acid. These acids combine, of course, with whatever bases or alkalies are present in the soil in which the nitrification takes place, usually lime or soda. This mineralisation of the ammonia is the last stage in the conversion of the unstable proteid substance into the stable and inoffensive mineral salts.

Nitrogen-fixing bacteria. The ultimate source of the nitrogen of the tissues of plants and animals is the atmospheric gas. But elementary nitrogen cannot be utilised by the vast majority of the organisms of either kingdom until it is combined with oxygen, and with some base to form nitrates or nitrites, or is in the form of some compound of ammonia: then the plants can utilise it; or until it is combined with oxygen, hydrogen, carbon, sulphur and phosphorus to form proteid: then the animals can utilise it. If then a considerable proportion of the combined nitrogen of the tissues of plants and animals has been combined from the elementary state there must be some means of bringing about this synthesis. Now when we attempt to combine atmospheric nitrogen and oxygen in the laboratory the most powerful means are required—either the electric spark or furnace. Yet we find that this synthesis is carried out by the vital activity of micro-organisms.

It has been known for many years that the leguminous plants

can grow on unmanured soils and can produce a crop richer in nitrogen than any other known to us. Further Kuhn, in 1901, shewed that he had been able to raise good and increasing crops of rye on one field for twenty years, and that more combined nitrogen was taken from the soil than was precipitated on it from the atmosphere. Obviously then the fixation of the elementary gas must have been carried out in the soil. Berthelot shewed in 1885 that this fixation was probably due to the action of bacteria, and just ten years later Winogradsky shewed, in his well-known memoir, that such micro-organisms were actually present, and indeed were very widely distributed in the soil. Now the nitrogen fixation of the leguminous plants is affected by the activity of a bacterium, *Bacillus radicicola*, which infests the nodular, gall-like swellings to be found on the roots of the leguminous plants such as peas, beans and clover; and the association of the two organisms, the plant and the bacillus, is a symbiosis, that is a metabolic partnership; or a case of parasitism of the plant on the bacillary growth, whichever view we choose to take—both have been advocated. The plant furnishes the bacillus with a carbohydrate food—sugar; and the bacillus absorbs the atmospheric nitrogen, combining it with oxygen, and the nitric acid after being combined with some base is taken up by the plant. So much has been made out by a long series of researches, but Winogradsky shewed that there were other bacteria which lived independently in the soil, or in symbiotic relationship with other micro-organisms, and that these microbes also possessed the power of taking up the atmospheric nitrogen and combining it with oxygen. Winogradsky called the bacterium he discovered *Clostridium*. Soon afterwards Beijerinck followed with the discovery of another nitrogen-fixing bacterium which he called *Azotobacter*. So far such organisms were known only from the land but in 1895 and 1896 Keutner and Keding investigated the occurrence of both bacteria and found that they were present also in the sea[1].

[1] See Jost's *Plant Physiology*, Lecture XIX., for a general account of the biology of *Bacillus radicicola*, and for an account of Winogradsky's researches on *Clostridium*. Keutner's work is published in the *Wiss. Meeresunters.* of the Kiel Kommission, Bd. VIII. Abth. Kiel, 1905. He also gives a good account of the literature of *Clostridium* and *Azotobacter*. Keding's work is contained in Bd. IX. of the same memoirs.

Clostridium and *Azotobacter* occur in all land soils except moorlands, and they are also present in sand dunes, especially when "Strandpflanzen" are abundant. They occur in sea water, in fresh water, and in the mud and sand from the sea bottom. They are present on the surfaces of both fresh water and marine plankton and on the surfaces of sea-weeds, and such animals as starfishes and molluscs. They have been described from the North Sea, the Baltic, and the sea off the coasts of Africa and the Malay Peninsula, and from the Indian Ocean. Probably both species are quite universally distributed.

Clostridium is an anaerobe and *Azotobacter* is an aerobe. Again the former is a bacillus, while the latter is a coccus. They have this character in common, that they can take up atmospheric nitrogen and fix it, forming nitrous or nitric acids, which then combine with whatever bases are present. They can both be grown in food solutions containing absolutely no fixed nitrogen. Nevertheless in such solutions nitrites and nitrates accumulate.

Clostridium can live in a solution containing absolutely no free oxygen: indeed the presence of this gas is inimical to the activity of the organism. Winogradsky cultivated it in a solution over which passed a continual stream of free nitrogen. It requires, of course, a source of carbon and is somewhat fastidious in this respect, demanding such carbohydrates as cane-sugar, laevulose, dextrose, &c. It cannot assimilate the higher alcohols, starch, cellulose or the organic acids. It requires potassium and magnesium salts and a source of sulphur. As it grows acid is produced, and this must be neutralised by the addition of calcium carbonate to the nutritive solution, free acid being detrimental to the growth of the bacillus. The salinity of the solution does not matter greatly, thus it can grow in both fresh and sea water and in water containing up to 8 % of common salt. Though it is an anaerobic germ which is harmed by oxygen it can yet live in solutions containing this gas in the free condition, if along with it there are other bacteria which by taking up the oxygen can shield the *Clostridia* from the influence of the gas. This is a case of true symbiosis.

Azotobacter resembles *Clostridium* in its reaction towards nitrogen. But it is an aerobic bacterium and grows best in the presence of abundant oxygen. It can avail itself of a wide choice of carbon food—the simply constituted organic acids appear to be a very suitable source of carbon. Because of its simpler tastes and its reaction towards oxygen *Azotobacter* is probably of more significance in the sea as a nitrogen-fixing organism than is *Clostridium*. But it is important to note that between them these two microbes can live and grow in a great variety of conditions.

Marine plants, plankton and the surfaces of the higher animals thus afford a habitat for our two bacteria. Here they find their carbon food, for they are resident in a mass of mucus which affords the necessary nutriment. Their mineral salts they obtain from solution in sea water; their oxygen and nitrogen from the same source. The association or symbiosis between the marine plants and the bacteria is a helpful one from the point of view of either. The waste carbonaceous matter of the plants or animals provides the carbon food of the bacterium, and the nitrite or nitrate elaborated by the bacterium is a source of food for the plant.

Denitrifying bacteria. If, from the point of view of the experimental chemist, the fixation of free nitrogen is an operation of much difficulty so also is the reverse reaction, the decomposition of the oxygen compounds of this element. Yet again we find that just this decomposition can be effected by the activity of bacteria. Denitrifying bacteria have been known to occur on the land for some considerable time, but not until quite recently were they known in the sea. Brandt, in 1898, predicted their occurrence in the latter element from theoretical considerations, and in 1902 Baur[1] described two species from the Baltic, and almost simultaneously Gran[2] described several species from the North Sea off the coasts of Holland. Baur's species were *Bacterium lobatum* and *B. actinopelte*; and Gran's species were *B. repens*, *B. trivialis* and *B. henseni*.

Baur made use of a solution containing potassium hydrogen phosphate, magnesium sulphate and carbonate, ammonium sulphate, and salt in the proportion in which this occurs in sea water. He inoculated this solution with mud from the bottom of a marine

[1] Baur, "Ueber zwei denitrificenden Bakterien aus der Ostsee," *Wiss. Meeresunt. Kiel Komm.* Bd. VI. Abth. Kiel, 1902.

[2] Gran, "Studien über Meeresbakterien. I. Reduktion von Nitraten und Nitriten," *Bergens Museums Aarbog*, No. 10, 1901.

aquarium, and in a few days obtained an abundant growth of bacteria. This mixed culture was then inoculated in a solution containing fish-broth, peptone and potassium nitrite. Again an abundant bacterial growth formed and ultimately the food solution became quite free from nitrite. Gas was freely evolved and this was shewn to be free nitrogen. Baur found that his bacteria could grow in an infusion of the sea-weed *Fucus* and in mussel broth.

Gran found that while some of his species of denitrifying bacteria could utilise simple carbonaceous food-stuffs, others were more fastidious and demanded complex materials. He found also that his bacteria behaved differently towards nitrogen compounds. Some could reduce nitrate to nitrite, and the latter to ammonia, and this compound to free nitrogen. Others again carried on the reduction only so far as ammonia. All the species described by both investigators were aerobic germs, but some of Gran's bacteria could live and grow in the absence of oxygen. Thus, just as in the case of the nitrogen-fixing bacteria, the denitrifying ones can live under very varied conditions of nourishment, and this is of great importance, for we never find pure cultures of bacteria in nature, but always various species associated together. Now it may often happen that the waste products of one species may serve as the food-stuffs for others. Obviously these symbioses are of great significance.

Temperature and denitrification. Thus the denitrifying bacteria can reduce nitrates, nitrites, and ammonia and set free elementary nitrogen from these compounds. In a typical culture of these organisms the solution after standing for a few days begins to froth and evolve gas. This gas is nitrogen. By-and-by this frothing ceases and if the culture be then tested for nitrites or nitrates it is found that these compounds have entirely disappeared. All the combined nitrogen originally present in the culture has been decomposed and the element has been returned to the atmosphere. Now the length of time required for the completion of this change depends on the temperature; that is the activity of the microbes, to which the decomposition of the nitrogen compounds is due, varies with the temperature, just of course as we find to be the case with all forms of bacterial action. The investigation of the relation of temperature to rapidity of denitrification is of the very greatest importance from our point of view. In a series of observations on this point Baur obtained the following results:

For *Bacterium actinopelte* in mixed culture.

(1) Temperature 25° C. Denitrification began about 24 hours after inoculation and lasted for a period of time varying from 7 to 11 days. At the end of this time combined nitrogen had disappeared from the solution.

(2) Temperature 15° C. Denitrification began about 4 days after inoculation and lasted for about 27 days after which the solution was nitrite-free.

(3) Temperature 4°—5° C. Denitrification began after about 20 days and after 112 days the reaction was still incomplete, that is the tubes still frothed shewing that nitrogen was still being evolved.

(4) Temperature 0° C. During a period of 107 days there were no signs of denitrification.

For *Bacterium lobatum* in mixed culture.

This bacillus gave very similar results. It was however able to denitrify at a temperature of 0° C. but the amount of nitrogen evolved was extremely small.

The investigation of pure cultures of both species gave results very similar to those obtained for mixed cultures.

Thus the activity of the denitrifying bacteria investigated so far depends on the temperature. It is greatest at about 25° C. and it practically ceases at the temperature of freezing water.

Now the reactions of bacteria are of the very greatest significance in the circulation of carbon and nitrogen in nature. Resuming very briefly the statements of this chapter we find that there are in the sea:

(1) The ordinary putrefactive bacteria which break down dead proteid matter with the result that the nitrogen of the latter is finally dissipated in the form of ammonia.

(2) The ordinary fermentative bacteria which break down carbohydrates and fats, with the result that the carbon of the latter is dissipated in the form of carbon dioxide, and the hydrogen in the form of water.

(3) The nitrifying bacteria which are able to oxidise the ammonia resulting from putrefactive decomposition, with the result that the nitrogen of the latter compound appears as nitrite or nitrate.

(4) The nitrogen-fixing bacteria which can use the elementary nitrogen of the atmosphere as a source of food, converting this into proteid substance.

(5) The denitrifying bacteria which reverse the action of the nitrogen-fixing bacteria, and reduce nitrate to nitrite, the latter to ammonia, and ammonia to free nitrogen.

(6) The remarkable sulphur bacteria, the significance of which has not yet been sufficiently investigated.

In the next chapter I propose to discuss the effect of these various groups of micro-organisms in bringing about the circulation of food matter in the sea.

CHAPTER XII.

THE CIRCULATION OF NITROGEN.

CHEMICAL analysis shews that the animal and plant body is mainly built up from the four elements, nitrogen, carbon, hydrogen and oxygen. Added to these there are the metals, sodium, potassium and iron, and the non-metals, chlorine, sulphur and phosphorus. Calcium or silicon are also invariably present as the bases of calcareous or siliceous skeletons. All these, with some others, are indispensable constituents of the organic body, and in an exhaustive study of the cycle of matter from the living to the non-living phases, and *vice versa*, we should have to trace the course of each. But we are accustomed to regard nitrogen as the characteristic constituent of living substance and it will be sufficient to consider this element alone.

At any moment there exists, on the earth, a certain mass of nitrogen, combined with other elements, forming the living substance of animals and plants. Another fraction of the gas is present in the atmosphere, in the soil, and in water, in combination with other elements in the form of non-living organic and inorganic substances, which are either being broken down into simpler forms, or are being built up into living tissues. Yet a third mass is present in the elementary condition in the atmosphere. Roughly speaking there are about 4345 billions of tons of nitrogen in the latter form. How much is present in the shape of lifeless organic and inorganic compounds we do not know, and still less have we any idea of the mass of nitrogen which exists in the living tissues of animals and plants: certainly it must be only a small fraction of the incredibly great mass that is present in the atmosphere. But when we think of the almost infinite numbers of plants and animals that have existed on the earth since life first began, it is evident that the total quantity of nitrogen which has been contained in their tissues must have been inconceivably great. But there is no accumulation of this element in the crust of the earth, and the composition of the atmosphere has been much the same since very remote times. Therefore the nitrogen which is now present in the bodies of living plants and animals must have been transmitted from organism to organism throughout biological time.

Minute portions of organic matter receive their legacies of life, and growing, take up nitrogen, and the other constituents of their bodies, from the media in which they live. Having lived its allotted time the organism dies, and its constituent elements return to the dust. Or perhaps it may be devoured by another organism and then part of its substance becomes rebuilt up into that of the animal which has consumed it. And yet again this animal may be the prey of others. Thus there is a transmigration of the atoms of the material bodies of organisms through long series of animate and inanimate forms. But while there is an ultimate Nirvana in the transmigrations of souls, the atoms of nitrogen are bound on the wheel of change, and ceaselessly pass through living and non-living phases. Imperious Caesar dead and turned to clay does not at once fill a crevice in a wall, or stop a bunghole, but may have a place in the architecture of humbler creatures. Perhaps only after many transmutations through living organisms may the nitrogen and carbon of the organism reassume the inorganic phase. But sometime or other, they return to the earth or become dispersed in the sea or air. Dante, in his vision, saw the souls of sinners impelled by a furious wind, and it may be that the material atoms, the substrata of the souls, are carried far and wide in the currents of the atmosphere.

Even the substance of the body of the individual animal has but a transitory existence when compared with the normal span of life of the organism. The circulation of nitrogen and carbon takes place in this manner: certain food-stuffs, proteids, fats and carbohydrates are ingested and are broken down in the alimentary canal into somewhat simpler compounds. The proteids are converted into peptones, or other substances; the fats are dissociated; and the carbohydrates are hydrolysed to form sugars. All these products of digestion pass through the walls of the alimentary canal into the blood stream and ultimately reach the cells of the bodily tissues. There they become assimilated to form the living protoplasm, or perhaps they become lodged among the particles of the latter as stored-up food material. This is the constructive metabolism of the animal body. But very soon the living tissue, or its included food-materials, are oxidised, that is, burned up by the inspired oxygen, or otherwise broken down; and as a result of this destructive metabolism, the mechanical energy and psychical characters, which are the manifestations of the life-activity of the animal, are exhibited. The products of this destructive metabolism again pass into the blood stream, and are finally thrown out from the body as waste substances, or excretions. This elimination of the effete products of the body takes place in various organs, the lungs and kidneys in the mammal, from which carbon dioxide, water, and some nitrogenous residue, such as urea, are discharged; or in analogous structures in the lower animals. Thus life is maintained by the continual death of parts of the tissues of the living body; and just as continually the latter are renewed by the assimilation of food materials, which assuming the organic phase become part and parcel of the living organism.

We cannot speak so precisely of the elimination of waste substances from the plant organism. There the energy requirements are far less than in the case of the animal, and we find that constructive metabolism is immensely greater, in the plant, than destructive metabolism. Therefore during life the mass of a plant continually increases, until, as in the case of great trees, an enormous bulk of organised substance is formed. In these cases life continues throughout centuries and the body of the plant is always becoming greater because comparatively little organic matter is being shed. In the case of long-lived animals, such as some reptiles, the body ceases to grow after a certain age has been attained, and the mass of food assimilated is exactly balanced by the mass of waste substances eliminated. But there are no definite excretory organs in the plant, and it is only by the removal, from season to season, of such things as dead leaves, bark, resins, gums, &c., that excretions or waste substances are eliminated.

Sometime or other in the history of the organism, and in one of many forms, the reproductive process begins. Animals form eggs or buds, and plants form seeds, buds, &c. A minute portion of the living substance—the immortal germ-plasm—is handed down in some form or other as the offspring. From the point of view of the race the reproductive act is the object of the life of the individual, and being completed, death comes as a natural consummation; and the lifeless vesture, or body, is given over to the forces of dissipation. The bacteria which inhabit the surfaces of the body, and which during the life of the organism were restrained by the vital activities of the latter, now become irresponsible. Those processes which during life proceeded slowly and orderly now go on rapidly, and are quite uncontrolled. Soon the dead tissues of the plant or animal become resolved into a few simple chemical substances, chiefly mineral dust, water and carbonic acid gas.

Putrefactive decomposition. Apart from the activity of putrefactive and fermentation bacteria, no change, except the cessation of the production of energy, and the reproduction of new organisms, would take place upon the death of an organism. Even the proteids of the animal body, the most unstable of chemical substances, would persist unchanged for apparently an indefinite length of time. It has been demonstrated that there are unaltered proteids and fats in the substance of Egyptian mummies belonging to the early prehistoric period (about 4000 B.C.)[1]. In the exceedingly dry atmosphere of Egypt, and possibly because of the action of the baths of common salt (which was the embalming substance employed), bacterial action is very largely arrested and putrefaction does not occur. Extreme cold also inhibits the activity of putrefactive micro-organisms, and so we have found the bodies of mammoths, which probably lived prior to the time when the

[1] See Schmidt, "Chem. u. biol. Untersuch. von ägyptischen Mumienmaterial," *Zeitschr. allgem. Physiol.* Bd. VII. 2 and 3 Heft, 1907. It has been stated that haemoglobin the colouring matter of the blood, has been recognised; and even that the proteids specific to the human body could be identified by serum reactions. This, however, appears to be doubtful.

THE CIRCULATION OF NITROGEN

earliest Egyptian mummies were prepared, frozen in the soil of Northern Siberia, and with the flesh so "fresh" that it has been eaten by dogs. But these are quite exceptional conditions, and it may be stated quite generally that as soon as a plant or animal dies, the organic substance of its body is broken down by the action of putrefactive bacteria.

The precise nature and course of putrefactive decomposition under different conditions are not at all thoroughly investigated. The decomposition of fats and carbohydrates is however a comparatively simple process. The former substances are decomposed into the glycerine and fatty acids of which they are made up, and then each of these products is further broken down into simpler organic acids, carbon dioxide and water. Ultimately these bacterial decompositions or fermentations proceed so far that the whole fat or carbohydrate is resolved into the two compounds, water and carbon dioxide, with perhaps traces of other gases, such as methane. In the case of the proteids however, the process of putrefaction is much more complicated, and a multitude of highly complex chemical substances are elaborated. Among these are albumoses and peptones, then ptomaines and amido-acids, evil-smelling substances such as indol, skatol and phenol, and simpler compounds like sulphuretted hydrogen, methane, ammonia, volatile amines, carbon dioxide, &c. The precise order of appearance, and proportions of these products, vary with the external conditions of the putrefaction. Most of the substances mentioned are however unstable, and are soon resolved into much simpler compounds. There is little or no difference between the putrefaction products of the nitrogenous constituents of the bodies of animals and plants, and probably no difference at all in the nature of the decomposition of their fats and carbohydrates. Even the relatively resistant cellulose of the plant ultimately is broken down by fermentation bacteria as in the case of the carbohydrate of the animal body.

It is usually assumed that the changes which take place in organic matter undergoing putrefaction are due entirely to the activity of bacteria. There is, of course, little doubt that the characteristic organisms of a putrefying mass belong to this group of micro-organisms. If we allow a small piece of flesh to stand for several days in some water in a warm place it will be seen that the liquid soon becomes turbid, and if we examine it microscopically we will find that innumerable bacteria, cocci, and spirilla are present. But infusoria will also usually be present and one can generally increase the numbers of the latter organisms by adding a little sugar to the putrefying liquid. There is a succession of organisms in a mass of decomposing sewage: first bacteria, then flagellates, infusoria, turbellarian worms, nematodes, oligochaete worms, and dipterous larvae. Much of the organic matter of the sewage passes into the air in the form of the bodies of flies which have fed on the other organisms mentioned.

Modern sanitary science has availed itself of the power of decomposing organic matter possessed by bacteria and other micro-organisms. In the "septic" or "biological" methods of treating the sewage of great human communities, the offensive matter is broken down by bacterial action. In these installations the crude sewage, after being "screened" so as to remove the coarser solid matters, is led into the septic tanks. There are two types of the latter. In one air is excluded from the putrefying sewage, and the bacteria which flourish in the tanks are anaerobes, that is, species which can live in the absence of oxygen. In the other type oxygen is freely admitted. In crude sewage we have a liquid which already contains enormous numbers of bacteria (from 6 to 12 millions per c.c. in London sewage) and which also contains food matters for the bacteria in the shape of dissolved and suspended organic matters, broken down proteids, fats, carbohydrates, cellulose, &c. These food substances are utilised by the micro-organisms, both bacteria and infusoria, and the latter multiply and live upon the organic matter, which therefore rapidly disappears. Even after remaining in the septic tanks for 24 hours there is already a considerable diminution in the proportion of putrescible matters present in the sewage.

Nitrification. Thus a number of substances, unstable and capable of further decomposition, are continually being produced as the waste substances of the bodies of living organisms. These waste substances, or excretory products, reach the drains of human communities, or enter the subsoil water, and so pass into lakes or streams. In the sea, or in fresh waters, the waste products of organisms are eliminated from their bodies into the surrounding medium. Organisms, or parts of such, die and are cast off, are buried, or are simply deposited upon the land, and in all kinds of corners. All these substances begin to undergo putrefactive decomposition if (as is usually the case) the conditions are favourable. Some of the products of this breaking-down, water and carbon dioxide, are stable and undergo no further changes. Others however are still unstable. Such are the numerous products of the putrefactive decomposition of animal and plant proteids. Considering only the changes that take place upon the land we may say that the water of rivers and streams and that of the subsoil, contains a certain proportion of the products of nitrogenous putrefaction—which one may loosely term " organic matter."

This organic matter is now further acted upon by the nitrogen bacteria. Altogether a host of nitrogenous substances are produced from proteids and allied substances in the course of putrefactive decomposition. All these have the same fate, that of further breaking-down by micro-organisms, some of which carry on the resolution of amido-substances as far as the stage of ammonia, while other bacteria, in the presence of abundant oxygen, oxidise this ammonia to form nitrous and then nitric acids.

In the process of the septic purification of sewage the crude liquid is first of all treated in the open or closed tanks, and a partial decomposition of the proteid and other organic matter takes place. The effluent which issues from these tanks is a liquid, which although not inodorous is far less offensive than the crude sewage. But if incubated at a temperature of 38° C. it will still putrefy, shewing that organic matter is still present. The septic tank effluent is therefore led into the " contact beds," which are large rectangular, shallow pits, filled with coke, clinker, gravel, or other porous filtering materials. These filter beds are flooded with the effluent, and the latter is then allowed to stand for a certain period, 24 hours or more. If the contact beds are carefully constructed, and if the process is carefully and intelligently directed (which is not always the case), the effluent which results is a clear liquid, which contains very much less organic matter than the septic tank effluent, which does not smell, and which, if incubated at 38° C. does not putrefy. The putrescible substances originally present have been further broken down by the activity of the nitrifying bacteria, with the result that much of their nitrogen is converted into nitrous and nitric acids, their carbon and hydrogen passing into the forms of carbon dioxide and water. In this process of purification much oxygen is absorbed. The process is intermittent, each contact bed being filled with the effluent, and then emptied and allowed to remain so for some time. The filtering material remains covered by a scum of matter containing the bacteria, but every time the bed is emptied of liquid, air enters and fills up its interstices. It is the oxygen of this air which is utilised in the oxidation of the nitrogenous organic matter[1]. It is of course absorbed during the nitrification of the organic matter, but since the bed is used intermittently, the filtering material is enabled to renew its stock of oxygen in the intervals of its work.

Denitrification. The process of resolution of organic matter does not end even with this conversion of proteid, or the metabolic products of the latter, into nitrous and nitric acids. The reader will remember that bacteria are to be found, widely distributed both on the land and in the sea, which have the power of reducing nitric to nitrous acid, the latter to ammonia, and ammonia to free nitrogen. Now there is reason to believe that such denitrifying bacteria play a very significant part in the circulation of nitrogen in nature. If the processes taking place in the sewage purification tanks and contact beds be minutely studied it will be found that there is sometimes an actual disappearance of nitrogen from the effluent. If it were the case that all the organic matter were decomposed so that its nitrogen were oxidised to nitrous and nitric acids then we should expect to find just as much of this element in the final effluent as was present in the crude sewage. The form of combination only would be changed. But this is not always the case. Letts[2] has shewn that such an actual dis-

---

[1] See Letts, *Rept. Royal Commission on Sewage Disposal*, Rept. 3, Vol. II. (Cd. 1487, 1905). Other sections of this report also contain much information on the question of the purification of sewage matters.

[2] *Ibid.*

appearance of nitrogen may take place and we can explain this fact only by assuming that the gas is dispersed into the atmosphere, the compounds containing it being completely broken down by the denitrifying bacteria that are present in the sewage effluents.

Drainage of nitrogen from land to sea. Thus the substance that constitutes the living materials of the animal and plant organism is sometime or other resolved into simple inorganic compounds, or even into its chemical elements. In the active, rapidly metabolising animal this change is continually taking place, all the more quickly the smaller the organism is. In the sedentary plant the process is much slower but a certain amount of decomposition of organic matter takes place there also. Then when the animal or plant body dies the dispersive process takes place more rapidly. If denitrification occurs then the greater part of the body may be resolved into the intangible gases, nitrogen, carbon dioxide and water vapour, and may vanish into "thin air" just as completely as if the tissues of the organisms had been consumed by fire. All that would be left would be the mineral constituents of the skeleton, the lime or silica, which composed the bones or shells. Even these would ultimately be dispersed through nature. Perhaps a similitude of the skeleton of the animal might be preserved as a fossil; perhaps as an impression on the strata among which it becomes embedded. But even here, the material of the fossil would be entirely different from that which constituted the dead body. And there is no doubt that only an insignificant fraction of the animals and plants that have existed on the earth has persisted in the fossil form. Of the water and carbon dioxide which result from decomposition processes, some part escapes into the atmosphere, and another part enters the soil and is washed down by rain into streams and water-courses. So also with the nitrogen, and the nitrous and nitric acids: some part escapes into the air as the elementary gas. We cannot trace the further history of this portion which is "imprisoned in the viewless winds, and blown with restless violence round about the pendant world." But some of these substances enter the rivers and ultimately are carried down into the sea.

Not all the products of nitrogenous decomposition which enter the subsoil water or rivers reach the sea. On the land there are plants, and microscopic organisms feeding like plants, and these utilise some of the carbon dioxide, and nitrites and nitrates, as food-stuffs. So also with regard to the rivers and lakes into which land drainage enters. A certain fraction of the waste substances produced upon the land is thus almost immediately again made use of by terrestrial and fresh-water organisms, but a considerable quantity does also reach the sea.

How great is this quantity? Obviously we can only estimate its mass in an approximate degree. But we know what is about the volume of water conveyed annually to the sea by the great rivers of the earth; and we know, in some cases, what is the approximate quantity of nitrogen compounds contained in the unit volume of the water of some rivers; and therefore we can calculate the approximate mass of nitrogen which drains from the land into the sea. The Rhine discharges annually some 65,336 millions of cubic metres of water into the North Sea, and one litre of this water contains, on the average, from 2 to 3 milligrammes of nitrogen in the form of dissolved compounds[1]. Taking the lower estimate we find that the Rhine alone carries 130 millions of kilogrammes of nitrogen annually into the North Sea. If we take the volume of all the other rivers entering this area as double that of the Rhine we find that every year about 390 millions of kilogrammes, or 383,000 *tons of nitrogen are added to the North Sea*. Probably the volume of water carried down to the sea from all the rivers of the world is not less than about 300 times that of the Rhine. If we take it as such we find that every year some 39 *billions of grammes of nitrogen, or about* 38 *millions of tons*, are added to the oceans of the earth.

This dissolved material entering the sea is at once dispersed by the action of tides, winds and currents. The water near the land ought to contain a greater proportion of these products of land drainage than is found in truly oceanic areas. Nevertheless it is difficult to demonstrate by actual analysis that this is really

[1] See Brandt, "Stoffwechsel im Meeres," *Wiss. Meeresuntersuch. Kiel Komm.* Bd. IV. p. 230, 1899. The estimate is founded on Boussingault's and Hoppe-Seyler's analyses. Brandt gives the references to the latter work.

the case. We know of course that the coastal waters are less salt than those further out at sea, the cause of this lower salinity being that the sea there receives a large addition of fresh water. These coastal waters then ought to contain more dissolved nitrogen, and there is no doubt that this could be demonstrated if it were not the case that these compounds are utilised almost as rapidly as they are added to the sea.

Utilisation of land drainage by marine organisms. All round the coast there is, in the littoral and laminarian zones, a selvedge of plant life, in the form of the larger sea-weeds. Below a depth of about 20 fathoms or so these larger algae begin to thin out, and in relatively deep water, even while still within the limits of the continental shelf, they practically cease to exist. On the shallow sea bottom near the land other plant life is also relatively abundant. There are immense numbers of diatoms living in the sand and mud near low water mark; and the algae there are sometimes also covered with deposits of these organisms. Pelagic protophyta, diatoms, peridinians, &c., are also more abundant in the shallow water near the shore. We also find that the invertebrate and fish life is more abundant in this region. Now it is evident that the greater density of plant life near the land is directly due to the fact that there is a greater amount of the ultimate food materials, nitrogen compounds and carbon dioxide, there, than far away from the land. These plant organisms use the substances mentioned as food-stuffs, building up starch and proteid from the carbonic acid and nitrogenous drainage of the land. Probably very little, or perhaps none, of this ever reaches the central oceanic areas. These sea-regions are, no doubt, self-supporting, that is, the inorganic food-stuffs of the plants there are derived from the decomposition of the dead bodies of the organisms inhabiting these areas.

With this conversion of the nitrogenous land drainage into plant substance, the upward or constructive metabolism of this matter may be, for a time at least, arrested. If all the algae, or protophyta, which feed upon the waste substances washed down from the land were eaten by animals, then the nitrogen, which we have traced from the land down into the sea, would pass by such transmutations into higher forms of life—into other re-incarnations. But the larger sea-weeds are not eaten to a great extent by animals. Fishes, molluscs, and some other marine animals browse upon them, and man does indeed eat certain succulent algae. But the amount of the larger sea weeds thus disposed of is inconsiderable and there is no doubt that an immense quantity of marine algae, directly nourished upon the nitrogen washed down from the land, is not utilised by animals but simply dies and putrefies, sometimes on a very large scale. On the shores of Belfast Lough, for instance, there grow immense quantities of the sea-lettuce (*Ulva latissima*), which are nourished upon the luxuriant food materials afforded by the sewage of Belfast. Purification of this sewage by the septic methods is quite useless, so far as the avoidance of the nuisance is concerned, for the *Ulva* appears to prefer the nitrates of the septic effluent to the organic matter or ammonia of the crude sewage[1]. There are no animals which utilise the sea-weed as food and the result is that great masses of the *Ulva* are simply washed ashore to die and putrefy, becoming a serious nuisance, and even a danger to health.

Therefore when the nitrogenous drainage from the land is utilised by the larger sea-weeds its further organic progress may be interrupted, the element again passing into the inorganic phase. So also with a certain proportion of the diatoms and other protophyta which have utilised this dissolved food-stuff. There is no doubt that vast quantities of these organisms must die and fall to the sea bottom, there to putrefy. This must happen in the Antarctic Ocean, where there is an immense accumulation of deep sea ooze formed predominantly from the dead shells of diatoms. We must remember that at the bottoms of deep seas the temperature of the water is very low, and that bacterial action, and therefore putrefaction, may proceed very slowly. It may be the case then that a large proportion of the organic substance of the diatoms falling to the sea bottom is utilised as food by the abyssal animals living there. We must recognise however that some at least of the organic plant substance built up from the nitrites and nitrates washed down from the land, at once undergoes putrefaction,

[1] See Letts, *loc. cit.*

with the result that precisely the same nitrogenous waste substances are again formed.

Another fraction of the nitrogen compounds entering the sea begins an upward series of transformations. A small proportion of the larger sea-weeds which utilise these substances is, as we have seen, eaten by marine animals; and another, and much larger, proportion of the protophyta consuming these same things becomes the food of innumerable creatures. I have already stated that there is evidence that many marine invertebrates feed upon the dissolved carbon compounds present in sea water, and it is probably the case that these animals also utilise, as sources of food, some of the dissolved nitrogen compounds (though perhaps much less of these than of the carbon food-stuffs). But it is also the case that many molluscs, other invertebrates, and even fishes habitually feed upon diatoms. At once then some of this nitrogen utilised by the protophyta returns to the land, for the fishermen catch the cockles, mussels and oysters, and the mullets and sardines which have eaten the diatoms. Yet again we find that the ubiquitous copepods feed upon the peridinians and other holophytic and myxo-trophic protozoa, and a great host of marine animals feed upon the copepods. Thus the herrings, mackerel, shads, pilchards, sprats, and other fishes habitually eat the copepods and smaller crustacea, and these fishes are caught for the public markets, the herrings, mackerel, shads and pilchards to be used directly as human food, and the sprats to be made into sardines and anchovy paste. Again many fishes eat the diatom-feeding molluscs, thus the plaice and flounder do so to a very great extent, and other fishes also make use of molluscs as occasional food animals. Plaice and flounders are then caught and taken ashore as human food. Many other fishes in their juvenile phases eat the diatoms and other protophyta, and though the fishermen do not intentionally catch these little fishes, the latter are still necessary if we are to have big ones.

Further many large fishes like the whiting, turbot, skate, ray and cod eat smaller fishes, or eat invertebrata, and if we examine the food of these latter animals we will find that it is either smaller fishes which eat the protophyta directly, or it consists of molluscs, or other invertebrata, which directly or indirectly also utilise the diatoms, peridinians, and their congeners as sources of food. So we can easily construct series of animals each of which is the food of the one higher in the series. Thus:—

Diatoms⟶cockles⟶flounders⟶man;
Diatoms⟶oysters⟶man;
Peridinians⟶copepods⟶sprats⟶whiting⟶cod⟶man;
and so on.

Thus the nitrogen compounds which are produced on the land from the putrefactive decomposition of the dead bodies of animals and plants or the excretions of the animals, are washed down into the sea; and a fraction of the total mass of nitrogen so transported may again return to the land in the form of the bodies of useful marine animals which have fed, directly or indirectly, upon this nitrogenous drainage substance.

Production by marine plants. The plants are the intermediaries between the inorganic food salts of the sea and the organic proteid, fat, and carbohydrate which form the food-stuffs of the higher animals. Let us imagine that in an enclosed sea-area into which land drainage percolates, plant life were suddenly to cease to exist. It is almost certain that animal life would also become extinct in such a case. Two conditions would produce this result: (1) the constant addition of salts of nitrous and nitric acids and ammonia would by-and-by render the water poisonous to animals; and (2) there would be no production of organic from inorganic materials. Possibly a certain small proportion of the latter substances would be utilised by some of the lower invertebrata; but it would only be the nitrogenous compounds of the type of urea or extractive substances that could so be utilised by saprozoic animals; and there is no evidence that nitrates or ammonia salts could be used as food. The animals would feed upon each other, of course, but among them there would be a number such as the mollusca, or smaller crustacea, which are accustomed to eat the protophyta. With the extinction of the latter the lower invertebrata would miss their accustomed food, would cease to multiply, and would finally become extinct, their disappearance being hastened by the ravages of the larger predatory animals, such as the fishes. The latter would be the last survivors in our microcosmos, and in time, their food also disappearing, they too would die out. There would be a continual resolution of the organic substance of the animal body into carbon dioxide, water, and inorganic nitrogen salts, but these substances could no longer be utilised. For a time bacterial life would find all the conditions of well-being. Putrefactive micro-organisms would flourish on the dead bodies of the animals that had died of starvation; and then the nitrifying bacteria would attack the decomposition products of the putrefying proteids. But this process of nitrification would tend towards the final disappearance of all trace of organic nitrogenous substance, and since the bacteria require these, or the less highly oxidised nitrogen compounds, as sources of energy, they too would cease to find their conditions of life and would also cease to exist. Even those micro-organisms which are able to reduce nitrous and nitric acids to free nitrogen require some carbon compound, other than carbon dioxide, as a source of food; and since all such substance would gradually become oxidised to the latter compound, the nitrifying bacteria too would apparently suffer extinction. So far as we can see the complete disappearance of plant life from such a sea-area would result in its utter sterility.

If, in an enclosed sea-area, the addition of nitrogenous and carbonaceous drainage from the land was suddenly to cease plant and animal life would nevertheless continue indefinitely. For such an area of water would be self-supporting, the animals feeding ultimately upon the plants, and the plants utilising as food-stuffs the excretory products of the animals. Possibly there would be denitrifying bacteria present in this sea-area, and one might conclude that the effect of the activity of these organisms would be gradually to reduce the amount of available nitrogen compounds present in the water, by reducing the inorganic forms of these to free nitrogen, which would be given off to the atmosphere. But, supposing that no addition of nitrites, nitrates or ammonia salts occurs, then it is probable that a struggle for existence, that is for these food salts, would arise between the protophyta and denitrifying organisms, and the effect of the latter would be minimised. From what we know of the conditions of life in the sea, it appears reasonable to suppose that nitrogen-fixing bacteria would also be present. These would fix free nitrogen from the solution of the latter gas in the water, combining the element with oxygen so that it could be utilised by the plants. As the nitrogen in solution in the water became used up more would be dissolved from the atmosphere. Therefore the denitrifying and nitrogen-fixing bacteria would balance each other.

One can demonstrate this on a small scale by setting up a sea water aquarium. A capacious but rather shallow tank or jar may be filled with sea water and stocked by some marine animals. Sea anemones, fishes like the butter fish (*Centronotus*), limpets, small shore crabs, small plaice or flounders (but not dabs), shrimps, and other marine animals live well in such small aquaria. Some kind of alga should be introduced, but usually the water contains the zoospores of such and it may be found that algae will spring up in the tank even if not intentionally introduced. An air circulation should be set up, not so much with the object of supplying oxygen as for keeping the water in movement. If such an aquarium be successfully established a balance between the animal and plant life will be struck. If too many animals be introduced at first some of these may die, and, if unnoticed, may poison the water by forming objectionable putrefactive products. But in a successful experiment the waste products of the one category of organisms will provide the food-stuffs for the other. The carbon dioxide exhaled by the animals will be assimilated by the plants, and the oxygen given off by the latter will be inspired by the animals. The nitrogenous excretions of the animals will be nitrified by the bacteria present, and the products of these reactions will afford the food-materials utilised by the plants. A condition of equilibrium, varying from season to season will be established, and the tank will become a self-supporting sea-area on a small scale.

Transfer of nitrogen from sea to land. Thus inorganic compounds of nitrogen are formed on the land as the result of the activity of micro-organisms, which break down the excretory products of organisms, or the dead bodies of the latter. These inorganic nitrogen compounds are washed down into the sea as sewage matters and land drainage. There a certain quantity is utilised as food by plants, which again are utilised as food by animals. A certain quantity of the latter are caught by fishermen and taken ashore. A certain fraction of the nitrogen entering the sea from the land is therefore returned to it in the shape of economic marine products. How great is this fraction?

In any attempt to answer this question we become bogged in a statistical quagmire. I am conscious of our inability to give very approximate figures. But one must make what use is possible of existing data. In 1899 Ehrenbaum[1] made the first reliable estimate of the productivity of the North Sea fishing grounds, and considering all the data then accessible came to the conclusion that about 875 millions of kilogrammes of fish were landed annually from this area.

Now since the beginning of the International Fishery Investigations we are in possession of rather more accurate statistical information and I have already given (in Chap. IX.) an estimate of the approximate value of the North European sea-fisheries. In the year 1904 the total weight of fish landed from the North Sea was not less than 967 millions of kilogrammes, and that taken from the fishing grounds of Northern Europe was not less than 1571 millions of kilogrammes. It has been estimated that one kilo. of fish contains, on the average, about 119 grammes of proteid, and this latter quantity (basing the calculation on Playfair's empirical formula for proteid) contains about 19 grammes of nitrogen. Therefore the North Sea, in 1904, supplied to the land about 18 millions of kilos. of nitrogen; and the wider fishing area about 29 millions of kilos. Putting the same values in English units, the North Sea supplied about 18,000 tons of nitrogen; and the whole North European fishing area about 28,000 tons.

I would repeat the caution so frequently expressed in these pages, that the fishery statistics are not at all accurately collected; and that such figures as I give here can only be regarded as approximations to the truth. But it is generally admitted that the fishery statistics really underestimate the mass of economic animals taken from the sea. Probably then the above estimates are minimal ones. Probably also, a certain mass (how great we we do not know) of organic matter is taken from the sea, and is not even indicated in the fishery statistics. Considerable quantities of sea-weed are taken from the shore and put upon the land

[1] *Mittheilungen deutschen seefischerei Verein*, Bd. xv. 1899.

as manure; and fairly large quantities of starfishes[1] are also so taken. Storms and high tides must strand considerable masses of sea-weed, &c. which rots and decays. If, taking these circumstances into account, we estimate the annual depletion of the North Sea as about 25 millions of kilogrammes of nitrogen, I think the estimate might be an approximate one. To put it at 30 millions would probably be too high.

Again any figures that may be given for the addition of nitrogen compounds to the sea can only be roughly approximate ones. We cannot say, with much accuracy, what volume of fresh water enters the sea *via* the great rivers, though just such estimates are frequently made use of by the geologists. Neither can we claim very great accuracy for the determinations that have been made of the amount of dissolved nitrogen compounds in river water; though again it may be observed that these estimates are of corresponding value to those of the amount of suspended mineral matter, or dissolved saline matter, carried down to the sea in the rivers. Such data have been made use of in attempting to estimate the rate of detrition of the land, or the age of the crust of the earth, and, so far as I know, are generally accepted[2]. Perhaps we may safely assign certain limiting values for the addition of nitrogen to the North Sea; and the annual amount of 290 millions of kilogrammes appears to me to be such a limiting value. For not only are nitrogen compounds carried down to the sea in solution in river water, but they are also precipitated upon the surface from the atmosphere. Every flash of lightning, or other electric discharge in the air, causes some of the elementary nitrogen and oxygen to combine, forming oxides of nitrogen, which is carried down to the sea, as nitrous and nitric acids, in the rainfall. Unfortunately it is impossible to estimate the amount of nitrogen fixation that takes place in this manner[3]. Then we have seen that bacteria are present in the sea which possess the power of

---

[1] In some parts of England £1 per ton is paid for this produce.
[2] See Appendix, Palaeochemistry of the primitive ocean.
[3] The amount of nitrogen in the form of nitrates, nitrites and ammonia that is annually added to the land from the atmosphere has been variously estimated. According to the results of the Rothamsted experiments about 4·4 lbs. are annually deposited by rain on each acre of land. "Rain" includes snow, hail and dew. Other and different estimates have, however, been made.

combining elementary nitrogen with other elements to form proteid. Of the two known genera *Clostridium* is anaerobic, while *Azotobacter* is aerobic. Between them they can make use of a variety of carbonaceous food-materials. They appear to be widely distributed, and their life-conditions are such as, apparently, enable them to carry on the nitrogen-fixing function under diverse circumstances. The elementary gas is present in solution in sea water, and as fast as it is taken out of solution more can be dissolved from the atmosphere. We cannot, of course, attempt any estimate of the mass of nitrogen compounds thus added annually to the sea; but it is evident that it cannot be neglected.

Let us take 30 millions of kilogrammes as the mass of nitrogen that is annually *taken from* the North Sea. Probably this is a maximum estimate, being greater than the actual mass. Again let us take 390 millions of kilos. as the mass of nitrogen that is annually *added to* the North Sea. Probably this is a minimum estimate, being less than the mass really added. Therefore we find that:—

(1) About 30 millions of kilos. of nitrogen *at the most* are annually taken from the North Sea, whereas

(2) About 390 millions of kilos. of nitrogen, *at the least*, are annually added to the North Sea. So that, even if we make every allowance for errors in the deduction of these estimates, it is still apparently the case that much more nitrogen, in a form easily assimilable by plants, is added to this area than is taken from it. Probably ten times as much combined nitrogen enters the area than is taken from it. What becomes of the excess?

Destruction of nitrogen compounds in the sea. The North Sea is, of course, in open communication with the Atlantic and Norwegian Seas, and much of the nitrogen compounds which enter it may be widely dispersed throughout the larger area. But this, obviously, does not afford any explanation, for the reverse is also the case, many of the food-fishes captured in the North Sea being migratory animals that enter it from without. Plankton also may be transported into the smaller sea. And we must remember that the same influx of nitrogen compounds goes on all round the shores of the Atlantic. Of course the length of coast-line bears a smaller proportion to the surface of the Atlantic, than does the North Sea coast-line to the surface of that water mass. But we ought also to consider the catchment area of the rivers as well as the relations between ocean surface and coast-line. Apparently these considerations do not help towards a solution of our question. We know that many other substances enter the sea in greater amount than are taken from it. Salt is carried down to the sea in the rivers, but then salt is probably taken from the sea in the evaporated water vapour. Silica and lime are also added to the sea, but both substances are accumulated there: silica in the form of diatom ooze, sponge spicules, flints, and radiolarian ooze; and lime in the form of calcareous oozes, coral reefs, shelly sands and gravels, nullipore formations, &c. Potassium and phosphoric acid are also added, but the former element accumulates at the sea bottom as glauconite, while phosphoric acid accumulates in the form of phosphatic nodules. Magnesium, manganese, iron, &c. are also added to the sea but all these elements accumulate there in certain forms. No permanent deposits of nitrogenous materials are however known to occur, or to have occurred, in the sea bottom deposits. There are no insoluble inorganic nitrogen salts, and the insoluble organic nitrogenous substances that we do know cannot resist bacterial activity. There are deposits of nitrates on the land—the S. American beds of Chili saltpetre for instance, and the potassium nitrate accumulations elsewhere. But these substances can only accumulate in hot and dry climates. Their permanent disposition in the ocean is obviously an impossibility.

It has indeed been suggested that organic matter is being laid down at the sea bottom, forming a stratum that may at some future time become available for the nutriment of terrestrial plants. One can hardly say that this is not possible, for organic remains must sink to the bottom of the deep seas, and at the very low temperature which prevails there bacterial activity, and consequently putrefactive decomposition, may be practically arrested so that nitrogenous matters may accumulate. Nevertheless there is no direct evidence in favour of this purely *à priori* hypothesis. Natterer[1] again revived, in 1894, an older view of Schlössing to the

[1] *Denkschr. Akad. Wien*, Bd. LXI. 1894.

effect that nitrogen compounds were given off from the sea to the atmosphere. He shewed that animal metabolism led to an apparent predominance of ammonia salts in the sea, and nitrates upon the land. In some way or other ammonia was volatilised from the sea into the atmosphere. It is true that ammonia, if present in solution in the sea in considerable proportion, would be given off, to some extent, to the air. But it is present in such minute proportions that we must regard this as improbable. Again carbon dioxide also results from the metabolism of marine animals, and we should expect that the free ammonia in the sea water would be neutralised by it, forming a carbonate of ammonium which would not be a volatile substance. Altogether Natterer's hypothesis appears to be an improbable one.

Denitrification in the sea. There remains therefore the hypothesis that the excess of nitrogen compounds in the sea is destroyed by bacterial activity, the nitrogen being returned to the atmosphere as the elementary gas. This was first suggested by Brandt[1] in 1899 before it was definitely known that denitrifying micro-organisms existed in the sea. It was obviously impossible to suppose that a greater mass of such salts as nitrates of calcium, sodium or potassium could be added to the sea than was taken from it. Otherwise the sea water would in the course of time become so surcharged with these substances as to form a medium in which it is impossible that animals could live. Then it was known that bacteria capable of converting nitric into nitrous acid, nitrous acid into ammonia, and the latter compound into free nitrogen, existed upon the land. We have seen that Baur and Gran were able to isolate allied bacteria, capable of carrying out the same reactions, from sea water. Here then was a very probable explanation of the disappearance of the excess of nitrogen compounds carried down from the land to the ocean. These substances are reduced by micro-organisms, so that their nitrogen reappears as the free elementary gas. This diffuses into the sea water, and from the latter it is given off to the atmosphere.

Density of marine life and temperature. The presence of bacteria possessing this power of reducing the nitrogen compounds

[1] *Stoffwechsel im Meeres*, 1899, *loc. cit.*

suggested to Brandt a possible explanation for the apparently paradoxical distribution of life in the warm and cold seas. I have already stated the evidence in favour of the belief that the colder seas of the temperate and frigid zones are not less rich in life than the warmer seas of the subtropical and tropical regions. Putting this fact of distribution in such a way as to avoid any straining or exaggeration of the case, we may say that on the land, the density of life, particularly plant life, decreases very greatly as we pass from the equator towards the poles, so that while there is a luxuriant vegetation in the tropical and subtropical regions, there is but little plant life on the polar land, even where the latter is not covered with snow or ice. But this decrease in plant and animal life does not take place in the sea as we proceed north or south from the equator. Indeed the opposite appears to be the case.

If it is the case that denitrifying bacteria are universally present in the sea, and if the activity of these organisms is not exactly compensated by the activity of equally ubiquitous nitrogen-fixing bacteria, then it seems reasonable to suppose that a greater amount of destruction of nitrogen salts will take place in the warmer than in the colder seas. For it is clear that the activity of denitrifying bacteria increases with the temperature. In the tropical and subtropical sea-areas these micro-organisms are very active and a considerable mass of nitrates, nitrites, and ammonia must be reduced to elementary nitrogen. But in the cold polar and temperate seas the activity of these bacteria must be greatly restricted, or even arrested entirely, by the low temperature of the water, and not nearly so much of the nitrogen salts will be destroyed. Therefore the warmer sea-areas must be much less rich in the nitrogenous food-salts, which are necessary for the nutrition of the plants, than are the colder waters. To that extent then there must be a lesser production of organic substance in the equatorial, than in the polar sea-areas.

While discussing such problems as these one is impressed with the enormous gaps in our knowledge of the conditions of life in the sea. It is quite evident that Brandt's hypothesis cannot be the only one capable of explaining, provisionally at least, the facts of distribution indicated above. Whatever explanations may be given must take account of the distribution of the food-stuffs of marine plants by oceanic currents, or of the manner of utilisation of these under different physical conditions. We must remember that animal and plant life is more luxuriant in the warmer land regions than in the colder ones. There must therefore be a greater drainage into the sea of the products of decomposition and metabolism from the tropical and temperate land areas than from those in the polar regions. If we could shew that a large proportion of this material, originating on the land in the warmer countries, were transported by oceanic currents into the seas of the colder regions then it might be possible to account for the wealth of life in the latter areas, without assuming the activity of denitrifying bacteria. An hypothesis of such a nature has recently been formulated by Nathansohn[1]. It is known that there are considerable vertical movements of the water of the oceans, that is currents may well up from the sea bottom, and conversely the water of the surface layers may fall down to the bottom, as for instance, when it becomes strongly cooled and sinks by convection. Again, we have seen that planktonic organisms living near the surface of the sea die and their bodies fall to the bottom, there slowly to putrefy. Therefore there must be a greater proportion of inorganic nitrogen compounds and carbon dioxide in these bottom waters than at the surface; and by the continual precipitation of dead organic matter to the bottom, the surface layers must become impoverished of their dissolved nitrogen food-salts. The latter, and carbon dioxide, cannot be utilised at the sea bottom to a great extent because plant life is scarce, or absent there, owing to the deficiency, or complete absence, of light. Now if an upward current of water transports these food materials from the bottom to the surface, the waters of the latter layers at once become enriched and there will arise an increase in the production of plants. Nathansohn shews that such upwelling of bottom water occurs in many parts of the sea, and that wherever it does occur there is an increased abundance of plant life. Obviously this upwelling bottom water will be colder than that normally present at the surface of the sea into which it emerges. Nathansohn works out

---

[1] "Über die Bedeutung vertikaler Wasserbewegungen f. d. Produktion des Planktons," *Abhandl. Math.-Phys. Klasse königl. Sächs. Gesell. Wissenschaft.* Leipzig, Bd. xxix. 1906.

these ideas, and by means of them accounts not only for the greater luxuriance of plant life in the cold seas, but also for the occurrence of the well-known maximum of plant production in the sea during the spring months.

There is still another explanation of the anomalous distribution of plant life in the warm and cold seas. The reader will remember that Pütter[1] shewed that both animal and plant metabolism is much more intense in warm than in cold sea water. With increasing temperature the rapidity of respiration of a cold-blooded animal also increases, and it uses up more oxygen, and gives off more carbon dioxide. With increasing intensity of respiration there is an increase in the general metabolism of the organism, so that the same animal uses a greater mass of food-stuffs per hour when the water is warm than when it is cold. Therefore in the warmer seas the ordinary food-requirements of animals and plants are greater than in the cold seas.

We must distinguish between what Pütter calls the "Betriebsstoffwechsel," and "Baustoffwechsel" of an animal or plant. The former we may call the ordinary current metabolism of its body, the assimilation of food-material and excretion of waste products which must necessarily proceed if the organism is to continue to live. The latter is the structural metabolism, the setting aside of assimilated food-substance for the formation of new tissue material, of eggs, buds, generally of new individuals. The former is the anabolism and katabolism of the ordinary life-processes of the stationary organism; the latter is the anabolism of the growing plant or animal, or of the animal which is actively reproducing. It is the metabolism of growth and reproduction.

Now in warm and cold seas alike the reproductive phase has become a habit of all organisms, recurring at regular intervals of time, which are determined by physical conditions. Alike in warm and cold seas there is a population of animals and plants which is exactly adjusted to the amount of food-materials present. If the current metabolism of a cold-blooded animal or marine plant increases with the temperature it follows that a lesser mass of food-stuffs will suffice for the same population in a cold sea than in a warmer one. But if the organisms of the plankton of a warm

[1] *Stoffhaushalt des Meeres.*

sea-area require a greater proportion of food-stuffs for the maintenance of their ordinary life-processes, it may be the case that the structural metabolism suffers; the assimilated food-substances being restricted to the ordinary metabolism of the organism (thus passing through a " current account "), rather than to the structural metabolism (passing into a " sinking fund "). Therefore it seems reasonable to expect that fewer individuals would be generated in warm than in cold seas, that is to say, the production would be less.

The reader will see that the problems of the causes of the variations in density of animal and vegetable life in the sea are exceedingly complex ones. But it seems clear enough that certain principal factors are concerned in setting up these variations of density of life: (1) the movements of sea water, either horizontally as oceanic currents, or vertically, as it sinks to the bottom or rises to the surface. This circulation of the sea water must distribute the inorganic nitrogenous and carbonaceous food-salts, leading to an abundance of these in some parts of the sea and a scarcity in other parts. Marine life will, of course, vary according to the variations in abundance of these ultimate food-stuffs. (2) The marine bacteria, particularly those that act specially on nitrogen and its compounds, are also factors. The denitrifying microorganisms are capable of reducing the nitrogen salts to such a form that they can no longer be assimilated by the plants. On the other hand there are also bacteria in the sea which can utilise the free nitrogen which is dissolved in the water from the atmosphere, converting this into compounds which can serve as food-substances for the plants. (3) Finally the mass of life in any part of the sea must depend to some extent upon variations in physical conditions: on sunlight because the energy of this is utilised by the plants in the process of photo-synthesis; and upon temperature, since with the rise of the latter, ordinary metabolic processes become more wasteful.

Resuming the main facts elicited in our study of the circulation of nitrogen we find that the land areas are being depleted of this substance (1) because in the metabolism of organisms, and in the decomposition of the dead bodies of these, nitrogen salts are formed which are washed down into the sea in the water of rivers;

and (2) because a certain amount of denitrification must take place. On the other hand the land gains nitrogen, (1) by the transfer of organic material to it from the sea; and (2) by fixation of the elementary nitrogen of the atmosphere by electric discharges, and by the activity of the bacteria which are associated with certain plants. The sea is being depleted of its nitrogen (1) by the transfer of economic organic products to the land (the fisheries); and (2) by denitrification. It gains nitrogen, (1) from land drainage; and (2) by nitrogen fixation by electric discharges and nitrogen-combining bacteria.

It is not possible to set, in balance-sheet fashion, these receipts and outputs against each other. We know only very imperfectly what are the approximate masses of nitrogen compounds in circulation. We do know, however, that the sea is not becoming appreciably richer or poorer in nitrogen compounds. Probably it gives up to the land and atmosphere just as much of this element, or its compounds, as it receives from those parts of nature. We know also that the composition of the atmosphere has remained very approximately the same during long periods of time. It too must receive back as much nitrogen (from the reduction of compounds of this element by bacterial life) as it yields to the land and sea (from fixation by electric discharges, and nitrification by micro-organisms). But we must not assume that the composition of the sea and the atmosphere remains quite constant. Perhaps changes are proceeding which by-and-by may profoundly influence life both on sea and land. Such changes are, however, taking place very slowly indeed.

In the present state of our knowledge this is all that we can say. Quantitative methods of study of biological phenomena in the sea, such as I have attempted to illustrate in this book, have only recently been adopted, and are still too imperfectly developed to enable us to trace numerically the course of circulation of the ultimate food-substances of terrestrial and marine organisms

APPENDIX I.

THE CHEMISTRY OF THE PRIMITIVE OCEAN.

A. B. MACALLUM[1], in an extremely interesting paper, discusses this question from the point of view of the composition of the blood and tissue-fluids of marine and land animals. Originally the earth was a molten mass, and the temperature was so high that many compounds now in the solid state were then in the form of vapour, and with their elements dissociated from each other, as is still the case in the atmosphere of the sun. But as the earth cooled down, and as the crust became solid, the temperature of the atmosphere fell, and then many of these elements combined together to form compounds, such as chlorides, sulphates, carbonates and oxides of sodium, potassium, magnesium, calcium and others, and these compounds were precipitated on the thin, solid, but still very hot crust of the earth. The temperatures at which these combinations and precipitations took place were very variable and depended on the pressure of the atmosphere, which was then much greater than it is at present, but was continually falling. Probably when these compounds were first condensed, refusion of the thin crust of the earth would take place over the areas of precipitation and thus diffusion of the salts would take place over considerable areas of the crust. Solidification and refusion would probably take place many times before a permanently solid crust could be formed. When the temperature of the latter fell low enough the water vapour of the atmosphere would also be precipitated, but it would again be evaporated; and these condensations and revolatilisations must have taken place very many times before the first permanent seas would be formed. There would be places on the earth where the temperature of the crust was lower than

---

[1] "The Palaeochemistry of the Ocean," *Trans. Canadian Institute*, vol. VII. 1904.

elsewhere and here the first seas would be formed. The primitive crust would be thin and yielding, and as the interior was still liquid, whenever a considerable mass of water fell on it, it would become depressed, and in this way the first ocean beds would be formed.

The composition of these first oceans would be very different from that of the modern ones. At first, for a very short time, the water in them would be almost pure but soon salts would accumulate, (1) from the solution of the materials present in the original crust of the earth, and (2) from the solution of the salts present in the rocks of the dry land, and brought down in solution by the rivers. In the water of these first rivers the salts would be present in proportions depending on the relative amounts in which they were present in the rocks, and also on their relative solubilities. The composition of the sea would soon become very similar to that of the rivers. Potassium and sodium would probably be very nearly equal in amount; possibly magnesium would be more abundant than sodium, and calcium more abundant than either. Changes would soon take place in the composition of the water of the primitive sea, for mutual reactions between the various salts would take place.

By-and-by life appeared in the sea and then further changes would take place in the composition of the ocean salts by reason of the action of living organisms. Calcium would be precipitated, for this element would be secreted by organisms to form their limy skeletons, and when they died the skeletons would sink to the sea bottom and accumulate. Some magnesium would also be precipitated, for we find this element in association with lime as magnesian limestone. Potassium would be precipitated as glauconite, a mineral which is formed by organic action. The decomposing dead matter of organisms liberates sulphuretted hydrogen, which then combines with iron to form ferrous sulphide, and then the sulphur of this compound is oxidised to form sulphuric acid. This acid acts on finely divided clay and sets free colloidal silica, and the ferric hydrate present combines with this to form a silicate and this then reacts with potassium salts to form glauconite. In this manner and by other means the excess of potassium originally present in the sea was gradually eliminated from solution. Whenever life began to increase in the sea these reactions would take place, and thus the calcium and potassium originally present in excess would be reduced in amount. Sodium is not removed from the sea in any great amount. Small seas may dry up and their salt contents may pass out of solution, but this process cannot take place on a very large scale. Salt is also removed from the sea by the evaporation of water, but it comes back again in the rainfall. During all the past history of the earth salt has been dissolved out from the rocks and carried down to the sea in the rivers and so it has gradually been increasing in amount. Thus because of the mutual reactions of the salts themselves; because of the action of organisms; and because of the solvent action of the rivers on the rocks of the dry land, the composition of the salts present in the sea has continually been changing: lime and potassium have been decreasing; sodium has been increasing; and magnesium has also been increasing.

In many animals the liquid in the vascular system has much the same composition as sea water; in fact we may regard it as sea water with some organic matter super-added. In the blood plasma of vertebrate animals the relative proportions of the elements sodium, potassium, and calcium are strikingly similar to the relative proportions of these same elements in the sea water of the present day. But whereas there are 11·99 parts of magnesium in the sea to every 100 parts of sodium there is only 0·8 part of magnesium in the blood plasma to every 100 parts of sodium. If again we analyse the substance of the muscles, &c. of the vertebrate body we find that the relative proportions of the elements sodium, potassium, magnesium and calcium are very different from those present in the blood.

Now there is probably a relation between the composition of the salts of the blood and tissues of the body and that of the salts of the watery medium in which this has lived. The first organisms probably originated in the sea. They were unicellular organisms, neither plants nor animals, but organisms from which both kingdoms of life have originated. They lived bathed in water in which were dissolved salts in certain proportions, and continually these salts in the sea reacted on the living protoplasm of the organisms living there. Thus the protoplasm acquired modes of metabolism which were due to the reaction to its first environment. The relative proportions of the elements sodium, potassium, magnesium and calcium present in this original protoplasm became similar to the proportions in which those same elements existed in the seas in which life appeared.

By-and-by organisms became more complex. First of all they became multicellular, and then among other structures they acquired a circulatory system. At first the fluid in this circulatory system was in open communication with the water of the sea: it was in fact modified sea water. We still find animals, coelenterates, of this kind in the sea. Then the circulatory system became shut off from the outside world and formed a closed system of canals. When this closure took place the composition of the blood, as regards the relative proportions of the inorganic substances in it, was the same as that of the surrounding sea water. Now these evolutionary changes, first that of the multicellular animal, then the acquisition of an open, and then a closed blood vascular system, took a very long time and during this interval the composition of the sea was continually changing.

The reader should remember how very strong is the influence of heredity, and how slowly, in spite of evolutionary changes, the more fundamental of the characters of living organisms change. Thus protoplasm is a mixture of immensely complicated substances. Each of the proteids contained in it is a molecule-complex composed of a great number of "building stones," each of which is itself a complex molecule. It is almost infinitely variable in composition, as variable probably as the species of organisms. Yet in spite of this variability we find it everywhere essentially the same in general characters and reactions, though a number of different types of structure are conceivable. If the protoplasm of both animal and vegetable cells is the same, then it is probable (arguing in the approved Darwinian manner) that both animal and vegetable protoplasm is similar, in general composition, structure and reaction to the protoplasm of those ancient organisms from which both kingdoms originated. Again, the highly complex process of nuclear division is similar in both animal and vegetable cells, and this argues in favour of the continuity of structure and reactions of protoplasm all down through the ages.

Why again should vertebrates, molluscs and crustacea have skeletons composed of carbonate and phosphate of lime instead of (say) siliceous or clayey skeletons, or perhaps iron ones? Silicon and aluminium are more abundant in the earth's crust than calcium, and one can easily conceive of bones or shells in which the inorganic matrix is silica or some alumina compound, or some oxide or carbonate of iron. Hosts of organisms have siliceous skeletons (though none have a skeleton in which the earthy basis is aluminium or iron). Obviously the skeletons of the vertebrata contain lime as their inorganic basis because at the time when the "provertebrata" lived in the sea the water of the latter contained a large proportion of calcium salts in solution. For long ages these organisms secreted lime from the sea and so a "lime habit" of metabolism became established and the necessity of this element is indicated in some queer facts of mammalian physiology: in the necessity for the presence of a calcium salt in the coagulation of the blood, for instance. The ancestral vertebrate organisms became adjusted to an environment in which lime was a prominent factor, and heredity has stamped this adjustment on the metabolism of the living vertebrates, which secrete their skeletons in the manner that their ancestors in Silurian times did. Usually one thinks only of morphological and psychical characters as being hereditary, but after all these are only transitory features of the relatively unalterable "protoplasmic basis of life."

So just because the primitive unicellular organisms lived in a sea in which the salts in solution were present in certain proportions their living substance came to contain these salts in the same relative proportions as did the sea water. Heredity fixed this proportion of the elements sodium, potassium, calcium and magnesium in the protoplasm of the primitive organisms, and we see it to-day in the composition of the protoplasm of living animals. Then multicellular animals possessing a circulatory system containing a liquid which had the same relative composition as that of the surrounding medium were developed. By-and by the circulatory system, originally in connection with the sea, became closed off from the latter, and when this happened the inorganic constituents of the blood were similar, in respect of their relative proportions, to those of the sea of the time. But even after the circulation had been shut off from the sea, and long after the composition of the latter had changed, heredity maintained the composition of the blood. The proportions of the salts of sodium, potassium, calcium and magnesium in the blood of living vertebrate animals are therefore those in which these salts were contained in the sea in very remote geological times.

APPENDIX II.

COMPOSITION AND DISTRIBUTION OF DEEP-SEA DEPOSITS.

MURRAY and Irvine (*Proc. Roy. Soc. Edinburgh*, vol. XVII. page 85, 1891) give the following table shewing the estimated area, the mean depths of occurrence, and the percentage of lime in relation to deep-sea deposits.

Deposit	Area in square miles	Mean depth in fathoms	Mean percentage of $CaCO_3$
Oceanic deposits:			
Red Clay	50,289,600	2727	6·70
Radiolarian ooze	2,790,400	2894	4·01
Diatom ooze	10,420,600	1477	22·96
Globigerina ooze	47,752,500	1996	64·53
Pteropod ooze	887,100	1118	79·26
Terrigenous deposits:			
Coral sands and muds	3,219,800	710	86·41
Other terrigenous deposits	27,899,300	1016	19·20

APPENDIX III.

FOUR HENSEN NET HAULS IN THE IRISH SEA. (13 Nov. 1906.)

Organisms identified	Proceeding in a straight line west from the English coast to that of the Isle of Man. Hauls equidistant			
	I.	II.	III.	IV.
Volume in cubic cents.	4	1	2	2·5
Asterionella bleakleyi	+	+	+	+
Biddulphia mobiliensis	+	+	+	+
Chaetoceros constrictum	+	+	+	+
Chaetoceros debile	+	+	+	+
Chaetoceros decipiens	+	+	+	+
Chaetoceros teres	+	+	+	+
Coscinodiscus concinnus	+	+	+	+
Ditylium brightwelli	+	+	+	+
Eucampia zoodiacus	+	+	+	+
Melosira borreri	+	+	+	+
Rhizosolenia semispina	+	+	+	+
Rhizosolenia shrubsolei	+	+	+	+
Ceratium furca	+	+	+	+
Ceratium tripos	+	+	+	+
Ceratium fusus	+	+	+	+
Tintinnopsis campanula	+	—	—	—
Noctiluca miliaris	+	+	+	+
Pleurobrachia pileus	1	0	2	0
Sagitta bipunctata	+	+	+	+
Larval polychaeta	+	+	+	+
"Mitraria"	+	+	+	+
Calanus helgolandica	1	0	0	8
Pseudocalanus elongatus	+	+	+	+
Paracalanus parvus	+	+	+	+
Temora longicornis	6	0	0	0
Acartia clausi	+	+	+	+
Oithona similis	+	+	+	+
Oikopleura	+	+	+	+
Copepod nauplii	+	+	+	+

See A. Scott, *Ann. Report Lancashire Sea-Fisheries Laby. for* 1906. Symbols denoting the frequency of occurrence of the organisms are used by the author, but in view of Apstein's objections to the use of these it is best to use only the signs of presence or absence.

APPENDIX IV.

THE ACCURACY OF THE OBSERVATIONS.

(1) **Calculation of the error.** If we assume that no constant error is involved in the use of the quantitative net, and that the plankton is distributed with absolute uniformity, then the experimental error of the observations may be calculated by statistical methods. A number of such determinations have been made and are summarised by Schütt (in *Analytische Plankton-Studien*, Kiel and Leipzig, 1892). The most obvious manner of estimating the error is by making parallel hauls with two nets coupled together on the same line and lowered and hauled at the same time. Schütt's Table 7 gives the results of six such experiments made during the Plankton Expedition. These are:

Hauls	Volumes of the individual hauls in c.cs.	Differences of the individual hauls from the mean of each pair in c.cs.	Percentage differences
I.	11 / 14	± 1·5	± 12·00 %
II.	14·5 / 15·5	± 0·5	± 3·33 %
III.	11 / 10	± 0·5	± 4·76 %
IV.	21 / 22·5	± 0·75	± 3·45 %
V.	36 / 35	± 0·5	± 1·41 %
VI.	241 / 208·5	± 16·25	± 7·23 %

The average error is thus 5·4 %. Calculated by the method of least squares it is 6·8 %. The probable error is however 4·5 %.

The results of 54 test hauls made by Hensen are also quoted by Schütt in Table 6. In these trials the same net was hauled twice in succession in the same place, and as nearly as possible from the same depth, and with as short an interval as possible, on 40 occasions; three times in succession on two occasions; and eight times in succession on one occasion. The average error was 12·8 %; the error calculated by the method of least squares was 19·7 %; and the probable error was 13·3 %.

Schütt gives also (pages 77—80) an analysis of the catches made in the Sargasso Sea during the Plankton Expedition, 24 hauls were made and the volumes of the catches (after removal of the larger objects) were—in cubic centimetres: 3, 4·5, 2·5, 2, 2, 3·5, 2, 5, 2·5, 6·5, 4, 2·5, 3, 4, 2·5, 3, 3, 4·5, 4·5, 3·5, 3, 5, 2·5, 1·5.

These catches indicate a remarkable degree of uniformity of the plankton in this part of the Atlantic. The simple average catch is 3·33 c.c.

Now if we take the separate catches we find that their divergences from the average catch are:

− 0·3, + 1·2, − 0·8, − 1·3, − 1·3, + 0·2, − 1·3, + 1·7,
− 0·8, + 3·2, + 0·7, − 0·8, − 0·3, + 0·7, − 0·8, − 0·3,
− 0·3, + 1·2, + 1·2, + 0·2, − 0·3, + 1·7, − 0·8, − 1·8.

The above values represent the divergences in c.cs. from the simple average catch. But calculating the mean divergence from the formula

$$M = \pm \sqrt{\frac{S}{N-1}}.$$

(where M = divergence, S = sum of the squares of the individual divergences, and N = the number of observations) and we have 1·2 c.c. as the mean divergence of each catch from the average one, and 0·8 as the probable divergence.

Because of the small catches made the percentage divergences are rather high. The average is 29 %; the average from the method of least squares is 36 %; and the probable divergence is 24 %.

But this is the total error and it is made up as follows:

Experimental errors
{
(1) Due to the motion of the ship during observations;
(2) Due to the imperfection of the net,
(3) Due to loss of catch on filtration;
(4) Due to error in reading the volume of the catch;
}

and (5) the error due to the variation in the distribution of the plankton.

Now it is difficult to separate the error due to the variation in uniformity from the other errors. Absolute uniformity of the plankton does not of course exist in the sea. Hensen's postulate is usually misunderstood. What the Kiel school of planktologists maintain is that wherever the physical conditions (*Lebensbedingungen*) are uniform there is also an approximate uniformity in the nature and distribution of the plankton. So we find, as Cleve's charts shew, that a large sea area, like the North Atlantic, or the North Sea, is to be divided into a number of sub-areas, each of which is characterised by a peculiar plankton-facies. The error due to experiment can only be separated from those due to irregularity if we make test hauls in an area where the physical conditions are as similar as possible over the entire area. Now such physical uniformity does not exist over even comparatively small areas of inshore seas. In the North Sea, as the charts published in the *Bulletin des Résultats* of the International Fishery Investigations Bureau shew, there are very great differences in the salinity and temperature of the water, from place to place, and from time to time; and with these differences there must be differences in the plankton. In the Irish Sea we have a certain homogeneity in the physical conditions, but this homogeneity is due to the mechanical mixing of the water coming from outside the area, with that coming from the land. It is not a uniformity due to the common origin of the water. One would therefore expect to find a certain degree of irregularity of the plankton due to the fact that some species which were adapted to live well in either of the water sources are not adapted for life in the mixture of both, and therefore that some of them will die out. Still less would we expect to find an inshore area adapted for such test hauls. In the Sargasso Sea, however, we have a vast mass of water occupying the centre of the Gulf Stream cyclonic circulation. It is what Haeckel called a Halistatic area, and in it we should expect a uniform plankton. Over 2000 miles of ocean the temperature during the cruise varied only from 26°·2 to 25°·4 C., and the salinity from 36·2 to 37. Schütt estimated that the total mean divergence of the average catch of 32% was due partly to the lack of uniformity, and partly to errors of experiment. The latter he estimated at 20%. Therefore the mean variation of the plankton in the Sargasso Sea in August 1883 was about 16% more or less than the mean.

(2) **Details of the catches made during the German periodic North Sea cruises.** The following figures are taken from Apstein's report in the *Wiss. Meeresunt.* Bd. IX. Abth. Kiel, 1906.

North Sea, 1903. (Figures in brackets indicate the stations.)

BIDDULPHIA spp.

February: (1) 16,000, (2) 320,000, (3) 48,000.

May: (1) 880, (2) 8000, (3) 0, (4) 0, (5) 80, (6) 20,000, (7) 3100, (8) 24,000, (9) 60,000, (10) 20,000, (11) 0, (12) 0, (13) 0, (14) 0, (15) 0.

August: (1) 60,000, (2—15) 0.

November: (1) 160,000, (2) 16,000, (3) 0, (4) 400, (5) 0, (6) 234,000, (7) 848,000, (8) 1,053,000, (15) 91,840,000.

COSCINODISCUS spp.

February: (1) 203,900, (2) 132,000, (3) 120,000.

May: (1) 36,000, (2) 4800, (3) 13,440, (4) 614,000, (5) 187,000, (6) 149,600, (7) 161,100, (8) 54,400, (9) 256,400, (10) 152,880, (11) 90,400, (12) 57,200, (13) 10,040, (14) 11,800, (15) 169,920.

August: (1) 5600, (2) 3400, (3) 400, (4) 4400, (5) 16,900, (6) 24,320, (8) 26,400, (9) 64,000, (11) 128,000, (12) 0, (13) 4000, (14) 2160, (15) 66,080.

November: (1) 27,000, (2) 12,080, (3) 91,200, (4) 9540, (5) 21,600, (6) 468,300, (7) 971,200, (8) 854,000, (15) 1,280,000.

CHAETOCEROS spp.

February: (1) 7,200, (2) 1,336,000, (3) 2,566,400.

May: (1) 518,400, (2) 32,000, (3) 520,800, (4) 4,000,000, (5) 4,876,000, (6) 4,548,000, (7) 1,012,000, (8) 627,000, (9) 11,304,000, (10) 3,756,000, (11) 157,977,000, (12) 10,020,000, (13) 24,000, (14) 144,000, (15) 248,000.

August: (1) 400,000, (3) 40,000, (4) 0, (4a) 0, (5) 48,000, (6) 0, (8) 0, (9) 724,800, (11) 2,760,000, (12) 0, (13) 1,424,000, (14) 784,000, (15) 18,400,000.

November: (1) 733,000, (2) 2,060,000, (3) 559,200, (4) 26,000, (5) 252,000, (6) 429,680,000, (7) 384,240,000, (8) 443,830,000 (15) 1,760,000.

CERATIUM spp.

February: (1) 880,000, (2) 368,000, (3) 2,960,000.

May: (1) 3,839,200, (2) 448,000, (3) 1,212,000, (4) 866,000, (5) 2,385,200, (6) 6,706,000, (7) 6,768,480, (8) 2,476,000, (9) 8,338,000, (10) 2,950,800, (11) 1,210,000, (12) 1,260,400, (13) 640,000, (14) 1,308,000, (15) 4,192,000.

August: (1) 601,600, (3) 11,640,000, (4) 293,680, (4a) 1,480,000, (5) 8,648,000, (6) 20,178,000, (8) 8,168,000, (9) 32,880,000, (11) 6,320,000, (12) 5,200,000, (13) 2,368,000, (14) 5,440,000, (15) 8,160,000.

November: (1) 2,017,000, (2) 336,000, (3) 21,536,000, (4) 19,066,000, (5) 12,044,000, (6) 21,498,000, (7) 15,536,000, (8) 14,719,000, (15) 2,800,000

COPEPODS.

February: (1) 336,960, (2) 96,160, (3) 213,300.

May: (1) 488,680, (2) 489,720, (3) 598,780, (4) 654,800, (5) 523,520, (6) 629,580, (7) 546,320, (8) 697,680, (9) 713,580, (10) 742,300, (11) 908,560, (12) 294,080, (13) 135,680, (14) 205,280, (15) 764,640.

August: (1) 381,600, (2) 1,344,440, (3) 1,044,960, (4) 1,391,520, (5) 442,560, (6) 742,960, (7) 620,820, (9) 4,232,000, (11) 697,440, (12) 478,160, (13) 1,327,840, (14) 396,640, (15) 261,600.

November: (1) 316,140, (2) 198,000, (3) 266,480, (4) 251,840, (5) 182,520, (6) 471,120, (7) 566,000, (8) 682,560, (9) 1,545,600.

(3) **Hensen's Nordsee expedition of 1895: details of the catches.** The results of this expedition are valuable, more because they are illustrative of a method of research which promises to be of service in practical fishery investigation, than because of the precise deductions made by Hensen. I think it is now certain that the enumeration of the various species of fish eggs was inaccurate. In 1895 the characters of the various species of teleostean eggs were not known very exactly. The Kiel zoologists based their identifications of the cod, flounder, plaice, whiting, dab and long rough dab mainly on the diameters of the ova, and we know (see Heincke and Ehrenbaum, *Wiss. Meeresunt. Kiel Komm.* Bd. III. Abth. Helgoland, Helgoland, 1900; and *Nordisches Plankton*, Kiel and Leipzig, 1905) that in certain cases the sizes of the eggs do not afford a very reliable means of distinguishing between the species of some ova. The individual catches of the cod eggs obtained are however of much interest as indicating the limits of variability of distribution of fish eggs in the

North Sea during the spawning season in question. Remember that it is possible that some cod eggs may have been wrongly identified.

These individual catches are[1]:

Cruise I. No eggs or larvae were taken in 18 hauls. The positive results were 9, 24, 24, 6, 3, 15, 6, 6, 54, 12, 165, 21, 27, 60, 204, 21, 3, 3, 3, 36, 24, 6, 9, 48, 21, 3, 15, 21, 108, 69, 3, 36, 48, 12, 18, 9.

Cruise II. No eggs or larvae were taken in four hauls. The positive results were 3, 9, 12, 24, 15, 216, 222, 183, 96, 156, 168, 543, 360, 24, 27, 312, 84, 36, 57, 39, 24, 75, 51, 69, 54, 78, 132, 252, 462, 126, 162, 153, 45, 69, 111, 131, 48, 39, 18, 6, 3, 105, 516, 63, 27, 75, 15, 12.

Cruise III. No eggs or larvae were taken in thirteen hauls. The positive results were 3, 33, 83, 29, 140, 60, 98, 71, 45, 44, 19, 9, 19, 2, 3, 1, 8, 16, 79, 55, 17, 5, 8, 34, 22, 11, 25, 15, 20, 13, 7, 14, 7, 4, 9, 8, 19, 81, 68, 24, 47, 17, 3, 9, 46, 5.

The reader will find synoptic representations of these catches in the distribution charts and curves in the paper by Hensen and Apstein already quoted. The probable error is discussed by Hensen. Even a cursory glance at the figures given above, with a reference to the chart of the cruises and the positions of the stations, will I think convince the reader that it is not impossible to make a roughly approximate statement of the abundance of all fish eggs in the North Sea which are of the pelagic type; that is, their distribution is not so variable that an average cannot be made. It may be argued of course that Hensen missed certain spawning grounds, or failed to fish in places where a greater abundance of eggs than was found might have been present, or conversely that he fished in places where there was an unusual abundance of ova. But this is pure conjecture and is à priori improbable when the distribution of the eggs actually found is considered. The analysis of the catches by well-known statistical methods will shew that there is a certain probable error in the estimate of the number of eggs present per square metre, and so of course in the estimate of the total number of eggs in the whole North Sea. So we may say, as Hensen does, that the actual number present may be so much less, or so much greater than the number actually estimated.

Fish eggs have absolutely no powers of locomotion, and larvae practically none, when we consider the size of the area. They are distributed by the mixing of the water due to winds, tides and currents. Thus the zoologists of the Nordsee Expedition found that

[1] *Nordsee-Expedition*, Tab. VII. p. 40.

the eggs and larvae of *Luidia*, a bottom-living sea urchin, were distributed through a very large area of the North Sea, although *Luidia* is a deep sea animal and is found only in two or three parts of the North Sea (see Hensen's Taf. XVII.). Of course the animal may have a much wider distribution than we suspect, but we have no right to assume this merely to discount the bearing of the fact of its wide distribution on Hensen's general statement. The bottom fauna of the North Sea is in fact pretty well known and if *Luidia* were generally distributed over the entire area it would have been more frequently found.

Again the sand-eel and the flounder are both shallow water fishes, that is, predominantly so. If these fishes spawned in the shallow water off the coast, and if there were no means whereby their eggs became widely distributed, then we should expect to find them in a comparatively restricted area near the land. But the fishes in question are shallow water spawners, and yet we find their eggs over the greater part of the North Sea, a fact which seems to me to justify us in concluding that there are unlikely to be any very considerable segregations of pelagic eggs in the sea, and therefore that the method of the Nordsee Expedition was a valid one. Nevertheless we must not forget that the limits of error in the results are somewhat wide; and that in the application of these results this limit must have a certain relation to the conclusions that have to be made.

APPENDIX V.

THE CALCULATION OF THE COEFFICIENTS OF THE QUANTITATIVE PLANKTON NETS.

This is an extremely difficult and laborious operation and I can only barely indicate the manner in which the calculations are made. If we were to attach only the ring which forms the mouth of the quantitative net to a rope and haul it up from a depth of (say) twenty metres to the surface, then a column of water of twenty metres in height, and of a sectional area equal to that of the opening of the ring, would pass through the latter. But if, again, we were to attach a bag made of some impermeable material to the ring and then haul it up through the water none of the latter would pass through the fabric, and instead we should have a pressure on the walls of the latter.

This pressure would be determined by the rapidity with which the apparatus was hauled. It can be ascertained by the application of the well-known Torricellian Theorem and is

$$D = \sqrt{\frac{V^2}{2g}}$$

where V is the velocity with which the apparatus is hauled, and g is the acceleration of gravity.

But whenever we attach the permeable silk net to the ring the case becomes quite different. The pressure D obtained as above no longer exists. It is as if a water main had been tapped or was leaking: then the pressure within it falls off. The net is not impermeable and water issues from each of its pores. If we wish to find how much water passes through the net fabric at a known velocity of hauling then we must determine what is the mean pressure on it per unit of filtering area. This pressure varies from part to part of the net: it is greatest near the mouth (near the instreaming water, and least at the end of the net, furthest from the instream, for immediately water enters leakage begins. The mean pressure on the walls of the water can only be determined by experiment and laborious calculation.

Starting out from such considerations Hensen obtains a coefficient, which he calls ψ, for each grade of the silk bolting cloth used in the construction of the nets. In making a haul with a quantitative net we have to consider (1) the velocity of haul—say 0·5 metre per second; (2) the area of filtering surface; and (3) the area of the mouth opening. The velocity of haul can be made always (approximately) the same; if not a correction can be applied. Suppose that the area of the mouth of the net is 0·1 sq. metre, then the volume of plankton caught, or the number of organisms, must be multiplied by the coefficient ψ in order to "approximate to" the number of organisms, or the volume of plankton which would have been contained in a column of water of 0·1 sq. metre in sectional area.

The methods used by Hensen in order to find the mean pressure on the walls of the net, and so the value of the coefficient ψ, are to be found in the *Bestimmung des Planktons*, and in the *Methodik*. If the reader has sufficient mathematical attainments, and is able to master the difficult German of these memoirs, he will, I think, find that considerable reliance is to be placed on the value of these constants. Of course the result of the reduction of the catch is to give only an approximate value for the plankton contents of the unit column of water. But when the net is carefully made and used (the haul must really be a vertical one and not the oblique haul that may be regarded as ' vertical ") then the coefficient doubtless gives reliable results. The limits of accuracy were investigated by Lohmann by comparison of the calculated catch of a vertical net with the catch made by the use of a pump and hose-pipe.

It may be suggested that the value of ψ varies with the continued use of the net. But the nature of the fabric of the latter is such, and the construction of the net is so planned, that the value of the constant cannot vary much during the limited period that the apparatus can withstand work at sea without disablement.

APPENDIX VI.

CALCULATION OF THE AGE OF THE EARTH FROM THE MEASUREMENT OF THE DRAINAGE OF SODIUM FROM THE LAND INTO THE SEA.

JOLY has attempted to estimate the age of the ocean from a knowledge of the amount of sodium in the sea, and the amount which annually enters it from the land. He finds from Murray's tables that the amount of sodium annually entering the sea is about 157,270,000 tons, and that the amount contained in all the oceans of the world is about 14,151,000,000,000,000 tons. Therefore the age of the ocean is apparently about 90,000,000 years. But Joly assumes that the first oceans due to the condensation of water vapour from the primitive atmosphere contained about 14 % of the sodium now present in the seas. On the other hand the amount of sodium in the sea is probably higher than is shewn in Murray's tables: Joly puts it as 15,627,000,000,000,000 tons. Again 10 % of the sodium entering the sea from the land comes from the former in evaporated water and is simply returned to the sea—not dissolved from out of the rocks. These corrections indicate that the more probable age of the ocean is about 89,300,000 years.

The Rev. O. Fisher criticises this estimate, pointing out that salt is probably imprisoned in the sedimentary rocks and is again returned to the sea on the weathering of the latter. This pre-indicates a circulation of salt to and from the sea. The effect of this correction is to multiply Joly's estimate several times.

Dubois also points out that the amount of salt dissolved out from the rocks is only about one-quarter of that deduced by Murray. If this is the case then Joly's estimated age must be quadrupled.

The literature of these calculations is summarised in a paper by Macallum in *Trans. Canadian Institute*, Vol. VII. Pt. 3, p. 536, 1904.

APPENDIX VII.

BIBLIOGRAPHY.

The following list includes only the more important memoirs.

(1) General: Methods and Results.

JENKINS, J. T. "The methods and results of the German plankton investigations, with special reference to the Hensen Nets." *Transactions of the Liverpool Biological Society*, Vol. xv. 1901. (An admirable short account of the Hensen investigations.)

SCHÜTT, FR. "Analytische Plankton-Studien." Kiel and Leipzig, 1892. (General, but deals chiefly with the results of the Plankton Expedition.)

HENSEN, V. "Ueber die Bestimmung des Planktons oder des im Meere treibenden Materials an Pflanzen und Thieren." 5 *Ber. Kommission Wissenschaftlichen Untersuchung d. deutschen Mere*, Berlin, 1887. (The original account of the methods and the first results.)

HENSEN, V. "Methodik der Untersuchungen." *Ergebnisse Plankton Expedition*, Kiel, 1895. (Deals with the methods adopted in the Plankton Expedition.)

HENSEN, V. and APSTEIN, C. "Die Nordsee Expedition 1895 des deutschen Seefischerei Vereins." *Wissenschaftliche Meeresuntersuchungen, Kiel Kommission*, Bd. ii. Heft 2, Kiel, 1897. (This contains a very interesting summary of the methods, and deals also with the productivity of the sea.)

APSTEIN, C. "Das Süsswasserplankton," Kiel, 1896. (Accounts of quantitative investigations in the fresh-water lakes of Holstein.)

REIGHARD. "Methods of plankton investigation in their relation to practical problems." *Bulletin United States Fish Commission*, for 1897, 1898, pp. 169—175. (An appreciative account of the Hensen investigations.)

(2) Results.

BRANDT, K. "Ueber den Stoffwechsel im Meere." *Wissenschaftliche Meeresuntersuchungen, Kiel Kommission*, Bd. IV. Kiel, 1899. (There is an English translation in the *Annual Report* of the Smithsonian Institution for 1901 but unfortunately the footnotes are omitted.)

BRANDT, K. "Ueber den Stoffwechsel im Meere," 2 Abhandlung. *Wiss. Meeresunt. Kiel Komm.* Bd. VI. Abth. Kiel, 1902.

BRANDT, K. "On the production and the conditions of production in the sea." *Rapports et Procès-Verbaux, International Council for the Exploration of the Sea*, Vol. III. Appdix D, Copenhague, 1905. (A summary of the results contained in the two previous papers.)

BRANDT, K. "Beiträge zur Kenntniss der chemischen Zusammensetzung des Planktons." *Wiss. Meeresunters. Kiel Komm.* Bd. III. Heft 3, Kiel, 1898. (Chemical composition of the plankton.)

APSTEIN, C. "Plankton in Nord- und Ostsee auf den deutschen Terminfahrten." *Wiss. Meeresunt. Kiel Komm.* Bd. IX. Abth. Kiel, 1906. (This paper contains estimations of the density of the plankton in the North Sea and in the Baltic.)

RABEN, E. "Über quantitative Bestimmung von Stickstoffverbindungen im Meerwasser" &c. (with a further paper on the same subject in the same Heft) *Wiss. Meeresunt. Kiel Komm.* Bd. VIII. Abth. Kiel, 1905. (This paper and its appendix contain the results of estimations of the proportions of silica and nitrogen compounds in the North Sea and Baltic.)

PÜTTER, A. "Studien zur vergleichenden Physiologie des Stoffwechsels." *Abhandl. könig. Gesell. Wissensch. Göttingen*, Bd. VI. Nr. 1, 1908.

PÜTTER, A. "Die Ernährung der Wassertiere," "Der Stoffhaushalt des Meeres." *Zeitschr. allgem. Physiologie*, Bd. VII. 2 and 3 Heft, 1907.

(3) Critical.

HAECKEL, E. "Plankton-Studien." Jena, 1890. (English translation in the *United States Fish Commissioner's Report* for 1889–91, Washington, 1893.) (In part an attack on Hensen's methods.)

HENSEN, V. "Die Plankton-Expedition und Haeckel's Darwinismus." Kiel und Leipzig, 1891. (Hensen's reply to the "Plankton-Studien.")

BRANDT, K. "Haeckel's Ansichten über die Plankton-Expedition." *Schriften des naturwissenschaftlichen Vereins Schleswig-Holstein*, Bd. VIII. Heft 2, 1891. (Deals also with Haeckel's criticisms.)

KOFOID, C. A. "On some sources of error in the Plankton method." *Science* (N.S.), Vol. VI. pp. 829—831. (Defects of the vertical plankton net.)

LOHMANN, H. 'Ueber das Fischen mit Netzen aus Müllergaze Nr. 20 zu dem Zwecke quantitativer Untersuchungen des Auftriebs." *Wiss. Meeresunt.* Bd. v. Heft 2, Kiel, 1901. (Critical examination of the grounds of Kofoid's objections.)

LOHMANN, H. "Neue Untersuchungen über den Reichtum des Meeres an Plankton." *Wiss. Meeresunt. Kiel Komm.* Bd. VII. Abth. Kiel, 1903. (This is quite a fundamental piece of research. It deals critically with the methods of plankton fishing.)

LOHMANN, H. "Untersuchungen zur Feststellung des vollständigen Gehaltes des Meeres an Plankton." *Wiss. Meeresunt. Kiel Komm.* Bd. IX. Abth. Kiel, 1908. (A development of the subject of the above two papers.)

APSTEIN, C. "Die Schätzungsmethode in der Planktonforschung." *Wiss. Meeresunt. Kiel Komm.* Bd. VIII. Abth. Kiel, 1905. (Deals critically with the methods of estimating plankton catches.)

HERDMAN, W. A. Presidential address to Linnean Society of London, 1907.

HERDMAN, W. A., and SCOTT, A. Intensive study of the marine plankton around the south end of the Isle of Man. *Annual Report Lancashire Sea Fisheries Laboratory*, Liverpool, 1908.

(4) Hydrography.

The literature is very scattered. Newer results are contained in the *Bulletin des Résultats* of the International Fishery Investigation Council, and in the *Rapports et Procès-Verbaux* (especially Vol. IL.). Valuable papers on hydrographical results and methods are also contained in the *Publications de Circonstance* of the same body. These papers cannot be neglected as they contain the most recent hydrographical results. The English Blue-Books giving the results of the English and Scottish fishery investigations carried on under the international scheme are also important. Their dates and numbers are Cd. 2670, 1905, and Cd. 2612, 1905. *Petermann's Mittheilungen* contains papers and summaries of interest; so also the *Scottish*

Geographical Journal. There are papers of importance by MURRAY and IRVINE in the *Proceedings*, and in the *Transactions* of the Edinburgh Royal Society. The *Challenger Reports*, of course; especially the *Narrative* and *Summary* Volumes, and the Report on *Deep Sea Deposits*, and the volume on *Physics and Chemistry*. CLEVE'S *Phytoplankton of the Atlantic and its Tributaries*, Upsala, 1897, should certainly be seen if it is accessible. A very clear, though elementary, account of the chief results of Oceanography is contained in H. R. MILL's *Realm of Nature* (Murray, 1895). This is by far the best of the physiography text-books. KRÜMMEL'S *Handbuch der Oceanographie*, Stüttgart, 1907, is the most recent book on the subject.

Papers dealing with the life-histories of fishes are so numerous that they cannot be summarised. But the reader should see CUNNINGHAM'S *Marketable Marine Fishes* (Macmillan, 1896); and MACINTOSH and MASTERMAN'S *Life Histories of British Marine Food fishes*. Both these books are founded on original observations. MACINTOSH, *Resources of the Sea* is often quoted but the book is not impartial and treats of a special subject. Zoological text-books devote, as a rule, little space to the accounts of the habits or Bionomics of fishes or other marine animals, but see the volumes of the *Cambridge Natural History*.

INDEX OF SUBJECTS

Aberdeen trawlers, 241
Absorption, coefficient, 236
  of food-stuffs, 220
Acanthias, 82
Accumulator, 32
"Albatross," 22
Albumoses, 262
Alcyonaria, commensalism in, 227
Algae, commensa., 226
Alimentary canal, 233
Amino-acids, 262
Anabolism, 209, 296
Anatomy, comparative, 229
Antiseptics, 261
Appendicularians, water filtered by, 140
  plankton filters of, 139
Aquarium, equilibrium in, 288
Aräometers, 12
Arctic-neritic plankton, 154
Arctic Ocean plankton, 151
Arctic Seas, abundant life in, 202, 205
  abundant diatoms in, 204
Ascaris, 229
Ascidians, 67
Aspergillus, 257
Atlantic plankton, 152
Atmosphere, primitive, 299
  mass of nitrogen in, 273
Aurelia, 98
Auftrieb, 56
Azotobacter, 267

Bacilli, 255
  Cholera, 258, 260
  B. coli, 257
  marine, 258
  B. radicicole, 267
  Bacterium actinopelte, 271
    lobatum, 271
Bacteria
  aerobic, 255
  anaerobic, 255
  and antiseptics, 261
  cell walls, 225
  circulation of nitrogen and, 271

Bacteria (*cont.*)
  collection of, 253
  denitrifying, 269
  desulphurising, 265
  distribution of, 253
  fermentation, 263
  flagella of, 255
  intestinal, in sea, 257
  leguminous plants, and, 267
  luminous, 259
  marine, 257
    characters of, 258
    numbers of, 260
  metatrophic, 261
  multiplication of, 254
  nitrifying, 265
  nitrogen-fixing, 266–7
  oxygen absorbed by, 243
  paratrophic, 261
  pathogenic, 261
  prototrophic, 260
  putrefactive, 260
  saprogenic, 262
  saprophile, 262
  sulphur, 264
  Syracuse, in sea off, 211
  temperature, and, 255
  types of, 256
Balanus, 97
Baltic
  fish eggs in, 244
  colour of, 41
  water level of, 48
Bank water, 47; and herring fi hery, 246
Barentz Sea, 49
  temperature and fisheries, 247
Barnacles, 69
Baulk net, 111
Baustoffwechsel, 296
Beche-de-Mer, 102
Beggiatoa, 264
Belone, 248
Benthos, 58
Betriebstoffwechsel, 296

INDEX OF SUBJECTS

Biddulphia, density in North Sea, 161, 309
Bilharzia, 229
Birth rate of copepods, 187
Biscay, Bay of, trawling in, 198
Blood plasma, 301
Brazil current plankton, 151
British fishing grounds, 32
British plateau, 31
Brooke sounding lead, 5

Calanus finmarchicus, heliotropism of, 147
Calcium in metabolism, 303
Carbon assimilation, 224
  by bacteria, 261
Carbon compounds
  in sea, 231
  source of, 239
  estimation of, 216
  Naples Bay, in, 216, 217
Carbohydrates, fermentation of, 263
  fermentation products, 277
Carbonate of lime in sea, 212
Carbonic acid in sea, 212, 216
Carnivores in sea, 221
Carp-culture, 179
Carp-ponds, 235
  productivity of, 184
Cellulose in animals, 225
  fermentation of, 277
Ceratium, 76, 89
  density in North Sea, 161, 309
  phosphorescence of, 259
Chaetoceros, density in North Sea, 161, 309
Chaeto-plankton, 151
"Challenger," 1, 10, 22, 56, 75, 77, 80, 213
Chemical Composition
  cod, 190
  copepods, 189, 192
  crab, 190
  diatoms, 189, 192
  flounder, 190
  herring, 190
  lobster, 190
  meadow hay, 190
  mussel, 190
  ox, 190
  oyster, 190
  peas, 190
  peridinians, 189, 192
  pig, 190
  plankton, 189
  potatoes, 190
  rye, 190
  salmon, 190
  sheep, 190
Chitin in bacteria, 225
Chiton, 72

Chlamydomonadeae, 226
Cholera bacillus, 260
Cilia, 86
Circulation in sea, 295
Climate and hydrography, 52
  and metabolism, 240
Clostridium, 267
Cockle, food of, 230
Cockling in Morecambe Bay, 114
Cod, 107, 248
  chemical composition, 190
  density in North Sea, 171
  fisheries in W. Baltic, 200
  influence of man on fishery for, 199
  spawning of, 247
  and temperature, 241
Cod eggs, density in North Sea, 170
  distribution in North Sea, 311
Coefficient of Hensen nets, 313
Collecting apparatus, 16
Colour of sea and plankton, 205
Commensal algae, 226
Commensalism, 226
Compound organisms, 226
Concinnus-plankton, 154
Consumers, 207
Continental shelf, 30
Contour lines of depth, 31
Convection currents, 143
Convoluta, 226
Conway mussel fishery, 183
Copepods
  birth rate of, 187
  characters of, 70
  chemical composition, 189
  density in North Sea, 161, 310
  in Baltic, 164
  destruction in Baltic, 188
  eaten by herring, 188
  food, as human, 191
  heliotropism of, 146
  length of life, 188
  production of in Baltic, 188
Coralline zone, 54, 59
  fisheries of, 60
Coral muds and sands, lime in, 304
  area of, 304
Corals, commensalism in, 228
  distribution with temperature, 241
Coscinodiscus, 79
  density in North Sea, 161, 309
Crab, chemical composition of, 190
Crust of earth, 299
  primitive, 299
Cucumaria, food of, 231
  metabolism of, 232
  respiration of, 232

Dab, density in North Sea, 171
  long rough, density in North Sea, 171
Davis Straits, plankton of, 153

J. F.

Deep-sea Deposits, 33
  composition, 304
  diatom ooze, 36
  distribution, 304
  extra-terrestrial, 37
  globigerina ooze, 35
  neritic, 34
  pelagic, 34
  pteropod ooze, 36
  pumice in, 34
  radiolarian ooze, 36
  red clay, 36
Deep-sea Fauna, 60
  distribution, 64
  food of, 62
  pressure and, 63
Denitrification, 280
  in sea, 293
  and temperature, 270
Depreciation of fishing grounds, 198
Depth-indicator, 3
Desmo-plankton, 151
Destruction in sea, 89
Development
  crustacea, 85
  direct, 84
  echinoderms, 86
  indirect, 84
  marine mammals, 82
  mollusca, 86
  oviparous fishes, 83
  phases of, 88
  plaice, 84
  sea-urchin, 245
  temperature, and, 243
  viviparous fishes, 83
Diatoms, 77
  Actinocyclus, 80
  Asterionella, 80
  Bacillaria, 80
  Bellerochea, 80
  Biddulphia, 78
  chemical composition, 189
  Coscinodiscus, 78
  density in North Sea and Baltic, 163
  Fragillaria, 80
  Gomphonema, 78
  Kiel Bay, in, 163, 236
  Navicula, 78
  Nitzchia, 80
  Pleurosigma, 78
  reproduction of, 78
  Rhizosolenia, 80
  spring maximum of, 236
  Thallasiosira, 80
  typical forms of, 79
Diatom ooze, 30
  area of, 304
  lime in, 30
Didymus-plankton, 154
Diffusion and salinity, 245

Drainage, nitrogenous, 281
Dredge, 16
Drift-net, 110
Dynamical equilibrium, 206
Dynamometer, 22

Earth, age of, 315
East Icelandic stream, 49
Echinoderms, 73
Edible Sea-urchins, 102
Enzymes, 225
European stream, 45
  Barentz Sea, in, 49
  distribution of, 46
  herring fishery, and, 246
  Norwegian branch of, 49
  offshoots of, 47
  origin of, 45
  plankton of, 153
  salinity of, 45
  temperature of, 45
Exploitation of sea, 196

Faeroe Islands, 31
  banks, 31
  channel, 31
Farm animals, chemical composition of, 190
Fats, fermentation of, 263
Fecundity of marine animals, 90
Fermentation, 224, 232, 263
  energy of, 263
  processes, 263
Filtrator, 125
Fishes
  chemical composition, 190
  density on sea floor, 175
  food of, 285
  immature, 108
  marking experiments, 175
  migration and temperature, 241
  nests of, 93
  numbers landed from North Sea, 174
  oviparous, 83
  phosphorescent, 259
  prime, 108
  protection of progeny, 93
  ratio of sexes, 172
  succession of, 99
  viviparous, 82
Fisheries
  benthic, 105, 114
  British, 181
  cockle, 114
  European, 181
  herring, 110, 245
  line, 109
  lobster, 114
  mackerel, 110
  mussel, 115
  North Sea, 180

INDEX OF SUBJECTS

Fisheries (*cont.*)
  oyster, 115
  pelagic, 105
  periwinkle, 115
  prawn, 114
  shell-fish, 114
  shrimp, 58
  sprat, 110
Fish eggs, 67
  Baltic, in, 244
  demersal, 83
  density of, 167
  destruction of, 92
  pelagic, 83
  salinity, and, 244
  succession of, 95
Fishermen, economic condition, 112
Fishery legislation, 196
  statistics, 181, 245
Fishing baskets, 23
  lines, 25
Fishing weirs, 25
Flounder, chemical composition, 190
  density in North Sea, 171
  eggs in North Sea, 171
Food of marine organisms, 221
Food-stuffs in sea, 223
Foreshore, 32
  fisheries of, 58
  organisms of, 58

Gephyrea, food of, 230
German plankton work, 309
Gill net, 111
Gill rakers, 222
Glauconite, 292, 300
Globigerina, 76
Globigerina ooze, 35
  area of, 304
  lime in, 304
Gordius, nutrition of, 229
Grafilla, nutrition of, 229
Greenland-Shetland banks, 31
  fjords, fauna of, 204
  seas, life in, 204
Gulf of Guinea, plankton, 242
Gulf stream, 38, 43, 308
  eddy, 45
  periodic variations, 50
  pulsations, 46
  unperiodic variations, 50

Haddock, 107
  density in North Sea, 171
Haemoglobin in Egyptian mummy, 276
Hake, 107, 248
Halibut, 248
Halibacteria, 258
Halistatic areas, 308
  plankton, 156

Halosphaera, 76
  and depth, 249
  plankton, 154
Harvest of sea, 196
Heliotropism, 248
  of *Calanus*, 147
  of copepods, 146
  of *Euchilota pilosella*, 147
  of *Obelia*, 147
  of *Phyllodoce*, 146
  of *Tomopteris*, 147
Hensen plankton net, 121
  accuracy of, 306
  calculation of error of, 307
  comparative hauls of, 306
  defects of, 141, 166
  experimental errors, 144, 307
  filtration capacity, 126, 314
  form for fresh water, 136
  validity of use of, 134
Herbivores in sea, 222
Heredity of chemical composition of animal body, 303
Herring, chemical composition of, 190
  fisheries, 110
  food of, 188
  migrations of, 245
  salinity, and, 246
  Skagerak, in, 246
  temperature, and, 246
  winter, 248
Hibernation in marine animals, 242
Holophytic organisms, 223
Holozoic organisms, 221
Homologous structures, 229
Homosaline sea areas, 308
Homothermic sea areas, 143
Hose nets, 115
Hydra, 226
Hydrographic charts, 30
  methods, 2
  soundings, 38
Hydrography and metabolism, 240
Hydrometers, 8

Iceland banks, 31
  cod spawning round, 247
  fisheries, 247
    and temperature, 247
  trawling, 198
Immature fishes in North Sea, 172
Impoverishment of sea, 196
Indispensable food-stuffs, 234
Infusoria and putrefaction, 278
Inorganic food-stuffs, 221, 232
Inshore fisheries, 106
Inshore plankton, 148
International fishery investigations, 7, 12, 27, 28, 50, 214, 246, 289
  area of, 15
  central laboratory, 13

INDEX OF SUBJECTS

Irish Sea, 33
  depths of, 33
  Hensen net, hauls in, 305
  homogeneity of, 308
  plankton of, 305
  salinity of, 241
Irminger current, 247
Isotherms in Atlantic, 38

"John Fell," 20, 124, 125

Karajakafjord, 204
Katabolism, 209, 296
Kiel Bay, plankton of, 205
  silica in, 236
Korbnetz, 127

Labrador current, 52
  fisheries of, 202
  origin of, 43
  plankton of, 153
  temperature of, 52
Labrax lupus, 241
Lake of Geneva, transparency of, 41
Laminaria, 33
Laminarian zone, 54, 59
  fisheries of, 59
  organisms of, 59
Lancashire cockle fisheries, productivity of, 183
Land crops, average, 193
Land drainage, nitrogen in, 282
  crops, chemical composition, 190
Larva
  copepod, 71
  crab, 70, 71
  fish, 68, 85
  megalopa, 70, 86
  mussel, 72
  nauplius, 71, 85
  pluteus, 68, 86
  trochosphere, 87
  veliger, 86
  zoea, 70, 85
Leguminous plants, 267
Leptocephalus, 65
Lime, secretion of, 241
Line fisheries, 247
  in North Sea, 109
Ling, 109, 248
Lithothamnion, 244
Littoral faunas, 235
Littoral zone 54
  fisheries of, 57
  organisms of, 57
Lobster, chemical composition, 190
  fisheries for, 105
Loch Fyne, 30
  Morar, 30
Lofoten cod fisheries, 202
Long-lines, 109

Lucas sounding machine, 4
Luidia, distribution of, 312
Luminosity of sea, causes of, 259
Luminous bacteria, 258, 259

Macrocystis pyrifera, 203
Magnesium in blood, 301
Mammoth, Siberian, 261, 277
Manganese deposits, 292
Marine bacteria, general characters, 258
Marine produce, chemical composition, 190
Mean squares, method of, 307
Mediterranean, currents of, 46
  nitrogen compounds in, 214
  plankton of, 165
  temperature of, 46
  thermal barrier of, 46
  transparency of, 41
Medusae, 74
Membrane, semi-permeable, 245
Metabolism
  analytic, 209
  animal type, 275
  ascidians, in, 233
  bacteria, in, 260, 261
  constructive, 209, 275
  current, 296
  destructive, 209, 275
  fishes and temperature, in, 242
  holothurians, in, 231
  intensity and temperature, 296
  molluscs, 233
  sponges, 231
  plant type, 275
  structural, 296
  Suberites, in, 231
  surface of organisms, and, 220
  synthetic, 209
Metre-wheel, 124
Micrococci, 254
Micro-plankton, 165
Microspora, 259
Migrations of herrings, 51
Migratory animals, 64
Migratory fishes, 65
Mineral salts in sea, 213
Minimum, law of the, 234
Molluscs, 72
  food of, 222
Morecambe Bay, 32, 58
  productivity of, 182
  mussel fisheries, 183
    productivity of, 185
  mussel transplantation in, 185
Müllergaze, 123
Mullet, food of, 222
Mummies, Egyptian, proteids in, 276
Mussel, chemical composition, 190
Mussel beds, 58

INDEX OF SUBJECTS

Mussel beds (*cont.*)
  culture, 184
  density of, 176
  transplantation, 185
Mytilidae, 102
Mytilus larvae in Baltic, 167
Myxosporidia, 225
Myxotrophic animals, 225

Nansen-Pettersson water bottle, 7
Naples Bay plankton, 204
  carbon compounds in, 217
  nitrogen compounds in, 217

"Nasse," 23
National Expedition, 135
Navicula, 230
Nekton, 56
Neritic deposits, 34
Neritic plankton, 154, 156
Nitrogen compounds
  added to sea, 291
  in Baltic, 215
  density of plankton, and, 235
  destroyed in sea, 291
  estimation of, 214
  Holstein lakes, in, 235
  law of minimum, and, 236
  Naples Bay, in, 215
  North Sea, in, 215
  produced by electric discharges, 290
  sea, in, 212
Nitrogen in atmosphere, 273
  circulation of, 274
  taken from N. European Seas, 289
  taken from North Sea, 289
Nitrogenous land drainage, 235
  excretions, 223, 275
Nitre-beds, 265
Nitre-deposits, 292
Nitrification, 266, 278
Nitrifying bacteria, 265
Nitrobacter, 265
Nitrococcus, 265
Nitrosomonas, 265
Noctiluca, 75
  phosphorescence of, 259
Nordsee Expedition of 1895, 168, 199
  cod eggs taken in, 310
  methods of, 169
North Atlantic, circulation of, 42
  isotherms in, 52
North Cape current, 49
Northern neritic plankton, 154
North Europe, climate of, 53
North fishes, 248
North Sea, 33
  colour of, 41
  circulation in, 48
  depth of, 33

North Sea (*cont.*)
  European stream in, 47
  fish eggs in, 167
  fisheries of, 107
  line fisheries, 247
  plankton of, 154, 159, 309
  productivity of, 180
  temperature of, 48
  transparency of, 205
  trawling in, 197
  under currents in, 48
  water-level of, 48
North-Western fishing grounds, productivity of, 181
Nuclear division, 302
Nutrition, modes of, 221

Ocean, age of, 315
  bed, primitive, 300
  deposits, area of, 304
Oceanographical apparatus, 2
  methods, 2
  research, 2
Offshore plankton, 148
Oikopleura albicans, 139
Organic matter, decomposition of, 279
Organised food-stuffs, 210
Ova of dogfishes, 66
  skates and rays, 66
Oxygen in sea, 213, 219
  Naples Bay, in, 219
Oyster, chemical composition, 190
  culture, 179
  fisheries, 115

Pacific ocean, transparency of, 41
Palaeochemistry of sea, 299
Pandalus, 108
Parasites, internal, 228
Pearl oysters, 102
Pecten, eyes of, 250
Penicillium in sea, 257
Peptones, 262
Peridinians
  chemical composition, 189
  density in Baltic, 164
  density in North Sea, 164
  phosphoric acid, and, 238
Petersen-Hensen net, 128
Phosphatic nodules, 292
Phosphorescence in sea, 259
  due to *Ceratium*, 259
  due to *Noctiluca*, 259
Phosphoric acid in sea, 212, 217
  and law of minimum, 236
Photobacillus, 259
Photobacteria, 259
Photo-chemical reactions in *Suberites*, 250
Photosynthesis, 223, 226, 249
Phototropism, 248

21—3

326   INDEX OF SUBJECTS

Phyllodoce, heliotropism of, 146
   and light, 248
Physical conditions and plankton, 308
   and metabolism, 239
Pipe-fish, 93
Placenta, 82
Plaice, 84
   density in North Sea, 171
   development, 84
   length and weight, 174
   marking experiments, 198
Plankton, 56, 65
   algae of, 98
   amphipods of, 70
   appendicularians of, 67
   ascidians of, 67
   calendar of, 94
   Ceratium, 75
   chemical analysis of, 129, 131
   chemical composition of, 189
   ciliata of, 75
   cirripedes of, 69
   coelenterates of, 73
   collection by pumping, 138
   convection changes, 148
   copepods of, 70, 99
   crustacea of, 69
   ctenophores, 74
   decapods of, 70
   dilution methods, 132
   distribution of, 142
   diurnal variations, 146
   echinoderms of, 73, 97
   estimation of, 129
      by enumeration, 133
      by mass, 131
      by volume, 129
      by chemical analysis, 131
   feeders, 222
   filtering apparatus, 125
   filtration, loss of, 136
   fish eggs in, 95, 97
   fishes of, 66
   flagellates of, 75
   food-stuff, as, 210
   food value of, 211
   foraminifera of, 75
   Halosphaera, 98
   heliotropism of, 145
   heteropods of, 72
   infusoria of, 75
   isopods of, 70
   Lakes of Illinois, in, 135
   larvae of, 166
   loss due to preservation, 136
   maximum catch of, at Kiel, 210
   medusae of, 98
   mollusca of, 72
   paper-filters for, 136
   pasture, compared with, 192

Plankton (*cont.*)
   peridinians of, 75
   pteropods of, 72
   radiolaria of, 75
   schizopods of, 70
   siphonophores of, 74
   Squilla in, 98
   statistics of, in North Sea, 309
   stratification of, 145
   succession of, 96
   test hauls of, 143
   Tintinnoidea, 75
   variation with latitude, 155
   worms of, 72
Plankton Expedition, 135, 143, 204, 242, 257, 258
Plankton-types, 150
   distribution of, 152
Plants
   characters of, 225
   food-stuffs of, 234
   insectivorous, 234
   metabolism of, 224
   motility in, 225
Plasma of blood, salts of, 301
Pleurobrachia, 74
Polyzoa, reproduction of, 88
"Poseidon," 214, 218
Potassium, elimination from sea, 300
   in blood, 301
Prawning, 108
Producers, 207
   and consumers, 166
Production
   absolute, of sea, 187
   by plants, 286
   in sea, 235
   in sea areas, 187
Productivity
   Baltic, of, 194
   beef on land, 193
   carp ponds, 184, 193
   cockle beds, 193
   Conway mussel fisheries, 182, 194
   cultivated land, 193, 194
      mussel fisheries, 185
   European fisheries, 181
   German carp fisheries, 184, 193
   inshore fisheries, 181
   Lancashire cockle fisheries, 183
   Morecambe Bay, 182, 194
   mussel beds, 193
   North Sea fisheries, 180, 194
   North-Western fisheries, 181
   shell-fisheries, 182
Proteids, composition of, 208
   as food-stuff, 208
   decomposition of, 261, 262
   Egyptian mummy, in, 276
   metabolism of, 275
   putrefaction of, 261

INDEX OF SUBJECTS

Proteids (*cont.*)
 putrefactive products, 262, 277
Protista, 74
Protophyta, 74, 75
 and carbon compounds, 239
Protoplasm, 302
 salts of, 301
Protozoa, 74
 saprozoic, 228
Provertebrata, 302
Proximate food-stuffs, 208
Pteropod ooze, 35
 area of, 304
 lime in, 304
Pteropods, 72, 203
Ptomaines, 262
Putrefaction, 276
 end products of, 262
 energy of, 263
 of proteids, 261
Putrefactive bacteria, 260
Pycnogonids, food of, 230

Radiolaria, 75
Radiolaria, "Challenger," Haeckel on, 77
Radiolarian ooze, 36
 composition and distribution of, 304
Red clay, 36
 area of, 304
 lime in, 304
Red coral, 103
Reproduction, modes of, 88
Respiration
 anal, 234
 fishes, in, 242
 in marine organisms, 232
 intensity and temperature, 296
 plants, of, 224
Respiratory surface, 233
 organs of invertebrates, 229
 quotients, 232
Rhine, volume of, 282
 nitrogen in, 282
Rhizosolenia, 96, 98
 density in North Sea and Baltic, 164
Rhizostoma, 98
Rhytina, 104
Rockall Bank, 31
Rothamsted experiments, 190, 290

Saccharomyces in sea, 257
Sagitta, 68, 73
 heliotropism of, 146, 248
Salinity, 13
 Baltic, in, 42
 Equatorial stream, 43
 estimation of, 13
 Gulf stream, 44
 Irish Sea, 308
 metabolism, and, 244

Salinity (*cont.*)
 North Sea, 43
 Norwegian Sea, 43
 sub-tropical Atlantic, 42
Salmon, chemical composition of, 190
Samples, plankton etc., 157
Sand-eels, eggs of, in North Sea, 312
Saprogenic bacteria, 262
Saprophile bacteria, 262
Saprophytic organisms, 224
Saprophytic plants, 225
Saprozoic animals, 228
Sarcina in sea, 254, 257
Sardine, food of, 89, 222
Sargasso Sea, 45, 308
 bacteria in, 260
 plankton of, 204, 242, 307
 salinity of, 308
 temperature of, 308
Scaphopods, 72
Scherbrutnetz, 27
Scottish trawl fisheries, 198
Sea-anemones, 73
Sea-bottom deposits
 benthic, 33
 diatom ooze, 36
 earbones in, 37
 extra-terrestrial, 37
 food, as, 63
 globigerina ooze, 35
 neritic, 34
 otoliths in, 37
 pelagic, 34
 pteropod ooze, 35
 pumice in, 34
 radiolarian ooze, 36
 red clay, 36
 terrigenous, 33
Sea-bottom, putrefaction at, 284
Sea-urchin, development of, 245
Sea-water
 colour, 41
 colour and plankton, 251
 density, 12
 examination of, 13
 halogens of, 13
 phosphorescence of, 259
 primitive composition, 300
 salinity, 13
 specific gravity, 12
 titration of, 13
 transparency and plankton, 251
 transparency of, 41
Seasons, retardation of, 52
Seasonal changes of plankton, 93
Seasonal fisheries, 99
Seine-nets, 26, 110
Sewage
 disposal, Royal Commission on, 257
 bacteria in, 278
 purification, 278

Sewage (cont.)
  putrefaction of, 273
  organisms concerned in putrefaction of, 278
Shell-fish, bacteria in, 257
Shell-fisheries, 116
Shrimp fisheries, 173
"Siboga" Expedition, 22
Silica
  and law of minimum, 236
  diatoms, and, 236
  estimation of, in sea, 218
  in North Sea, 218
  in sea, 212, 217
Siliceous deposits, 35, 292
Siliceous skeletons, 212, 302
Sira-plankton, 151, etc.
Skagerak, herring in, 246
Skeletons of organisms, 213, 302
  calcareous, 213, 302
  siliceous, 213, 302
Smacks, 106
Smacksmen, 112
Sodium in blood, 301
  drainage to sea, 315
  in sea, 315
  increasing in sea, 301
  removed from sea, 292, 313
Sole, 248
  and temperature of sea, 241
Sondeur à clef, 5
Sounding leads, 5
  machines, 3
South-fishes, 248
Spawning habits, 243
  places, 169
Spirallae, 255
  in sea, 258
Spirochaetes, 255
Spitzbergen, vegetation of, 201
  marine life near, 202
Sponges, metabolism of, 232
  water filtered by, 231
  skeletons of, 35, 213
Stake-nets, 25, 111
Staphylococci, 254
Statistics, fishery, 289
Stempel-pipette, 133
Stickleback, 93
Streptococci, 254
  in sea, 257
Styli-plankton, 151
Suberites, food of, 231
  metabolism and light, 250
  respiration of, 232
Summer fishes, 248
Sunlight and metabolism, 248
Symbiosis, 226
Syracuse, plankton of, 165, 210

Taffeta silk, 139

Temperature of land, annual range, 205
Temperature of Sea
  annual range of, 205
  Atlantic, 37, 38
  Irish Sea, 38
  North Sea, 38
  Antarctic, in, 41
  currents, 147
  diurnal variations of, 147
  equatorial stream, in, 37
  metabolism, and, 241
  North Atlantic, in, 37, 39
  Norwegian Sea, in, 40
Terrigenous deposits, 33, 304
  area of, 304
Thermal barriers, 38, 46
Thermometers, insulating, 11
  Negretti and Zambra, 8
  reversing, 8, 11
  Richter's reversing, 8
  Siemen's electrical, 10
  slow reaction, 10
Tidal streams, and plankton, 143
Tierra del Fuego, marine life off, 202
Tow-net, surface, 26
Trammel-net, 25, 111
Transparency of sea, and plankton, 205
Trawl
  Agassiz, 23
  beam, 17
  Heligoland young fish, 27
  otter, 18, 28
  shrimp, 18
  use of, 18
Trawlers
  cart-shanking, 107
  deep-sea, 22
  in Iceland seas, 107
  in North Sea, 107
  sailing, 106
  steam, 105
Trawling, method of, 21
  North Sea, 107
  second-class, 106
  shrimp, 108
Tricho-plankton, 151
Trichodesmium, 80, 153
Tridacna, 241
Trilobites, 87
Tripos-plankton, 151
Tropics, marine life of, 201
  vegetation of, 201
Turbot, 248

Ultimate food-stuffs, 212
Ulva latissima, 284
Unicellular algae, 77
Uniformity of plankton, 143

INDEX OF SUBJECTS

Vascular system, 301
  evolution of, 302
  liquid in, 302
Vertical plankton nets, 121
  circulation in sea, 295
  movements of plankton, 148, 295
Viscosity of sea-water and movements of plankton, 148

Water sample bottles, 7
  and plankton collection, 138
West Baltic trawl-fisheries, 199
Whalebone whales, 203
  food of, 203
White Sea, 202
Whiting, 107

Wild Birds Protection Act, 101
Winter-fishes, 248
  herring, 246
Worms, 72
Wyville-Thomson Ridge, 31
  temperature of sea over, 38
  salinity of sea over, 40

Xanthellae, 226

Yeasts in sea, 257

Zoochlorellae, 226
Zoophytes, 73
Zoorema, 135

INDEX OF AUTHORS

Apstein, on
  cod eggs in Nordsee Expedition, 311
  density of plaice and cod in North Sea, 311
  estimation of plankton, 129
  Nordsee Expedition, 168
  plankton of Holstein Lakes, 235
  plankton of North Sea and Baltic, 159
  use of Hensen net 135

von Baer, on life in Arctic seas, 202
Baur, on denitrifying bacteria, 269
Baxter, on mussel transplantation, 185
Beijerinck, on nitrogen-fixing bacteria, 267
Bidder, on bottom trailers, 16
Biebahn, on productivity of land in beef, 193
Boussingault, on nitrogen compounds in river water, 282
Bowley, on marriage statistics and birth rates, 238
Brandt, on
  abundance of diatoms in Kiel Bay 163, 210
  carp ponds, 184
  chemical composition of marine animals, 190
  chemical composition of plankton, 189
  circulation of nitrogen, 294
  denitrification in the sea, 269
  law of minimum, 236
  life in tropics, 204
  nitrogen compounds and density of plankton, 235
  nitrogen compounds in Rhine, 282
  plankton of Kiel Bay, 205
  productivity of land and sea, 201
  silica and diatom abundance, 236
Breitfuss, on
  fisheries of Barentz Sea, 247
  salinity of Barentz Sea, 49

Cleve, on
  European stream, 53
  plankton types, 151
  salinity of Gulf Stream, 44
Coppen Jones, on bacteria, 254

Darwin, on
  fecundity of marine animals, 91
  life in seas off Tierra del Fuego, 202
Dohrn, on food of Pycnogonids, 230
Dubois, on age of earth, 315

Ehrenbaum, on
  characters of fish eggs, 310
  fishery statistics, 289
  " Scherbrutnetz," 27
Ekman, on salinity of Gulf Stream, 4=

Fischer, A., on
  bacteria and circulation of matter, 254
  nutrition of bacteria, 260
Fischer, B., on
  marine bacteria, 257
  number of bacteria in Sargasso Sea, 260
Fischer, E., on composition of proteins, 262
Fisher, Rev. O., on age of earth, 315
Fol, on transparency of sea, 41
Forbes, on zones of life, 54
Fulton, on
  depletion of sea, 198
  fecundity of fishes, 171
  herring fisheries, 110

Gamble, on commensalism in *Convoluta*, 226
Garstang, on
  calendar of plankton, 94
  depletion of sea, 198
Gough, on heliotropism of plankton, 147
Gran, on denitrification in sea, 269

INDEX OF AUTHORS

Haeckel, on
 crustacea compared with insects, 69
 Plankton-Studien, 56, 135
 radiolaria, 77
Heincke, on
 length and weight of plaice, 174
 north and south fishes, 248
 plaice marking experiments, 175
Henking, on
 length and weight of plaice, 174
 plaice marking experiments, 175
 winter and summer fishes, 248
Hensen, on
 "Bestimmung des Planktons," 56
 cod eggs in North Sea, 310
 coefficients of vertical plankton nets, 314
 comparison of plankton fishing experiments, 143
 density of fish eggs in North Sea, 167
 depreciation of Baltic cod and plaice fisheries, 200
 distribution of *Luidia*, 312
 enumeration of plankton organisms, 133
 his neighbours' income, 179
 Korbnetz, 128
 life in sea and land, 195
 Nordsee Expedition, 168
 Mytilus larvae in Baltic, 167
 plankton of Kiel Bay, 205
 productivity of Baltic in plankton, 194
 productivity of carp ponds, 184
 productivity of sea, 188
 use of vertical plankton nets, 121, 135
Herdman, on copepods as human food, 191
Herlet, on eyes of *Pecten*, 250
Hickson, on commensalism in corals, 227
Holt, on mature and immature plaice in North Sea, 172
Hoppe-Seyler, on nitrogen compounds in river water, 282

Irvine, on
 composition and distribution of deep-sea deposits, 304
 metabolism of diatoms and silicic acid, 220
 nitrogen compounds in the sea, 213, 217

Jenkins, on fishery statistics, 182
Joly, on age of oceans, 315
Jost, on plant physiology, 265

Keding, on nitrogen-fixing bacteria, 267

Keeble, on commensalism in *Convoluta*, 226
Keutner, on nitrogen-fixing bacteria, 267
Knudsen, on
 salinity of North Sea, 43
 winds and land temperatures, 53
Kofoid, on
 defects of Hensen nets, 135
 plankton collection by pumps, 138
Kröyer, on fauna of Arctic seas, 202
Krümmel, on
 carbonic acid in the sea, 216
 oxygen in the sea, 219
Kutscher, on luminous bacteria in human intestine, 258
Kyle, on temperature and herring fishery, 246

Leach, Miss M., on chemical composition of bacteria, 225
Letts, on
 putrefaction of sewage, 280
 Ulva latissima, 284
Liebig, on law of the minimum, 234
Lohmann, on
 appendicularians, 140
 comparison of methods of plankton collection, 141
 defects of Hensen nets, 135
 Kofoid, 136
 micro-plankton of Mediterranean, 165
 Oikopleura filters, 139
 plankton fishing by pumps, 138
 plankton methods, 137
 plankton of Mediterranean, 211
 stratification of plankton, 145

Macallum, on
 age of earth, 315
 palaeochemistry of ocean, 299
MacIntosh, on
 resources of the sea, 197
 succession of the plankton, 94
Mill, on salinity and temperature of Firth of Clyde, 143
Möbius, on North and South fishes, 248
Monaco, Prince of, on
 sounding lead, 5
 surface floats, 15
Moore, on development of plaice and sea-urchin, 245
Müller, on use of tow-net, 55
Murray, on
 composition and distribution of deep-sea deposits, 304
 metabolism of diatoms and silicic acid, 220
 nitrogen compounds in the sea, 213
 silicic acid in the sea, 217
 sodium chloride in the ocean, 315

Nansen, on current meters, 16
Nathansohn, on
 circulation of nitrogen compounds, 295
 vertical movements of plankton, 295
Natterer, on
 circulation of nitrogen, 292
 nitrogen compounds in the sea, 213, 214

Ostwald, on viscosity of sea water and plankton, 148

Pantanelli, on photosynthesis, 249
Patten, on eyes of *Pecten*, 250
Pettersson, on
 European stream, 53
 herring fishery and salinity, 247
 line fisheries and salinity, 247
 salinity of Gulf Stream, 44
Plimmer, on composition of proteids, 262
Pratt, Miss E., on commensalism in Alcyonaria, 227
Pütter, on, carbon compounds in Naples Bay, 217
 dissolved food-stuffs in the sea, 234
 intensity of metabolism and surface area of organisms, 220
 Lithothamnion, 244
 metabolism of bacteria and oxygen, 243
 metabolism of *Suberites* and *Cucumaria*, 231
 nitrogen compounds in Naples Bay, 213, 216
 origin of carbon compounds in sea, 239
 photochemical reactions in sponges, 250
 respiration of *Suberites* and *Cucumaria*, 232
 structural and current metabolism, 296

Raben, on
 nitrogen compounds in the sea, 213, 214
 phosphates in the sea, 217
Rauschenplatz, on food of marine organisms, 230
Regnard, on life under great pressure, 64
Reibisch, on development of fish eggs and temperature, 243

Richard, on fishing baskets, 23
Roaf, on development of plaice and sea-urchin, 245
Rodewald, on productivity of the land in beef, 193
Ross, Sir John, on the "deep-sea clam" 55

Sarasin, on transparency of sea water, 41
Schimper, on vegetation of tropical and temperate shores, 202
Schlössing, on circulation of nitrogen, 292
Schmidt, C., on phosphates in sea-water, 217
Schmidt, J., on
 cod spawning and temperature, 247
 migrations of the eel, 64
 fisheries of Iceland, 247
Schmidt, W. A., on proteids in Egyptian mummies, 276
Scott, A., on
 Irish Sea plankton, 95, 305
 mussel transplantation, 185
Scott, T., on depths of Scottish lochs, 30
Schütt, on
 accuracy of plankton observations, 306
 colour transparency of sea and plankton, 205, 251
 plankton of Sargasso Sea, 204
 plankton of cold seas, 242
Shipley, on food of Gephyrea, 230
Smith, Miss K., on salinity of North Sea, 43
von Steman, on productivity of carp ponds, 184

Tanner, on deep-sea dredging and trawling, 22
Thompson, D'Arcy, on weight and length of plaice, 174

Vanhöffen, on plankton of Greenland seas, 204

Whiteley, on development of plaice and sea-urchin, 245
Winogradsky, on nitrogen bacteria, 265, 267

See also names of Authors, and works, under BIBLIOGRAPHY, p. 316.

HISTORY OF ECOLOGY
An Arno Press Collection

Abbe, Cleveland. **A First Report on the Relations Between Climates and Crops.** 1905

Adams, Charles C. **Guide to the Study of Animal Ecology.** 1913

American Plant Ecology, 1897-1917. 1977

Browne, Charles A[lbert]. **A Source Book of Agricultural Chemistry.** 1944

Buffon, [Georges-Louis Leclerc]. **Selections from Natural History, General and Particular, 1780-1785.** Two volumes. 1977

Chapman, Royal N. **Animal Ecology.** 1931

Clements, Frederic E[dward], John E. Weaver and Herbert C. Hanson. **Plant Competition.** 1929

Clements, Frederic Edward. **Research Methods in Ecology.** 1905

Conard, Henry S. **The Background of Plant Ecology.** 1951

Derham, W[illiam]. **Physico-Theology.** 1716

Drude, Oscar. **Handbuch der Pflanzengeographie.** 1890

Early Marine Ecology. 1977

Ecological Investigations of Stephen Alfred Forbes. 1977

Ecological Phytogeography in the Nineteenth Century. 1977

Ecological Studies on Insect Parasitism. 1977

Espinas, Alfred [Victor]. **Des Sociétés Animales.** 1878

Fernow, B[ernhard] E., M. W. Harrington, Cleveland Abbe and George E. Curtis. **Forest Influences.** 1893

Forbes, Edw[ard] and Robert Godwin-Austen. **The Natural History of the European Seas.** 1859

Forbush, Edward H[owe] and Charles H. Fernald. **The Gypsy Moth.** 1896

Forel, F[rançois] A[lphonse]. **La Faune Profonde Des Lacs Suisses.** 1884

Forel, F[rançois] A[lphonse]. **Handbuch der Seenkunde.** 1901

Henfrey, Arthur. **The Vegetation of Europe, Its Conditions and Causes.** 1852

Herrick, Francis Hobart. **Natural History of the American Lobster.** 1911

History of American Ecology. 1977

Howard, L[eland] O[ssian] and W[illiam] F. Fiske. **The Importation into the United States of the Parasites of the Gipsy Moth and the Brown-Tail Moth.** 1911

Humboldt, Al[exander von] and A[imé] Bonpland. **Essai sur la Géographie des Plantes.** 1807

Johnstone, James. **Conditions of Life in the Sea.** 1908

Judd, Sylvester D. **Birds of a Maryland Farm.** 1902

Kofoid, C[harles] A. **The Plankton of the Illinois River, 1894-1899.** 1903

Leeuwenhoek, Antony van. **The Select Works of Antony van Leeuwenhoek.** 1798-99/1807

Limnology in Wisconsin. 1977

Linnaeus, Carl. **Miscellaneous Tracts Relating to Natural History, Husbandry and Physick.** 1762

Linnaeus, Carl. **Select Dissertations from the Amoenitates Academicae.** 1781

Meyen, F[ranz] J[ulius] F. **Outlines of the Geography of Plants.** 1846

Mills, Harlow B. **A Century of Biological Research.** 1958

Müller, Hermann. **The Fertilisation of Flowers.** 1883

Murray, John. **Selections from *Report on the Scientific Results of the Voyage of H.M.S. Challenger During the Years 1872-76*.** 1895

Murray, John and Laurence Pullar. **Bathymetrical Survey of the Scottish Fresh-Water Lochs. Volume one.** 1910

Packard, A[lpheus] S. **The Cave Fauna of North America.** 1888

Pearl, Raymond. **The Biology of Population Growth.** 1925

Phytopathological Classics of the Eighteenth Century. 1977

Phytopathological Classics of the Nineteenth Century. 1977

Pound, Roscoe and Frederic E. Clements. **The Phytogeography of Nebraska.** 1900

Raunkiaer, Christen. **The Life Forms of Plants and Statistical Plant Geography.** 1934

Ray, John. **The Wisdom of God Manifested in the Works of the Creation.** 1717

Réaumur, René Antoine Ferchault de. **The Natural History of Ants.** 1926

Semper, Karl. **Animal Life As Affected by the Natural Conditions of Existence.** 1881

Shelford, Victor E. **Animal Communities in Temperate America.** 1937

Warming Eug[enius]. **Oecology of Plants.** 1909

Watson, Hewett Cottrell. **Selections from *Cybele Britannica*.** 1847/1859

Whetzel, Herbert Hice. **An Outline of the History of Phytopathology** 1918

Whittaker, Robert E. **Classification of Natural Communities.** 1962